PLANE

Practice Workbook with Answers

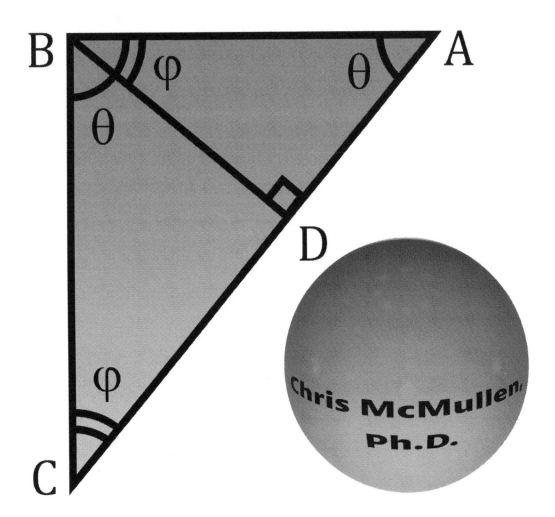

Plane Geometry Practice Workbook with Answers
Volume 1: Triangles, Quadrilaterals, and Other Polygons
Chris McMullen, Ph.D.

www.improveyourmathfluency.com
www.monkeyphysicsblog.wordpress.com
www.chrismcmullen.com

Zishka Publishing
ISBN: 978-1-941691-88-5

Mathematics > Geometry

Contents

Introduction

The goal of this workbook is to help students master essential geometry skills through explanations, examples, and practice.

- Each chapter begins with an introduction to the pertinent concepts and includes examples illustrating how the concepts can be applied.
- Several exercises in each chapter offer ample practice.
- The first 8 chapters focus on triangles. Why? Triangles illustrate fundamental concepts, such as congruence, similarity, and area. Also, since a common and practical method for working with other polygons is to divide the shape up into triangles, it is important to master the geometry of triangles before advancing onto quadrilaterals and other polygons.
- The book begins by introducing basic properties of lines, angles, and triangles in Chapters 1-2.
- Chapters 3-4 cover congruence and similarity, which have applications in many branches of geometry.
- Chapter 5 focuses on right triangles, including the Pythagorean theorem and special triangles.
- Chapter 6 discusses perimeter and area.
- Chapter 7 covers medians, bisectors, and altitudes.
- Chapter 8 introduces the triangle inequality.
- Chapter 9 covers quadrilaterals, including the square, rectangle, parallelogram, rhombus, trapezoid, and kite.
- Chapter 10 covers a variety of principles relating to polygons.
- Answer key. Practice makes permanent, but not necessarily perfect. Check the answers at the back of the book and strive to learn from any mistakes. This will help to ensure that practice makes perfect.

Lines and Angles

This chapter focuses on fundamental properties of lines and angles, which will serve as the building blocks for many other principles of geometry. Topics include parallel lines, perpendicular lines, complementary angles, supplementary angles, acute angles, right angles, obtuse angles, vertical angles, and the important case of a transversal that intersects a pair of parallel lines.

Chapter 1 Concepts

A **point** is a dot that has no size. A point represents a specific location in space. When a point is drawn in a diagram, it is drawn as a dot. While the dot appears to have size (so that the dot can be seen on the paper), the point that the dot represents does not have any size. A single uppercase letter (like A or B) is usually used to label a point in a diagram. The letter A was used to label the point below.

A
•

A **dimension** is a measure of extent. A point has zero dimensions because it does not extend in any directions; it has no size.

A **line segment** is a straight path that connects two points. The points at the end of a line segment are called **endpoints**. A line segment is finite. A line segment has a single dimension because it extends only in one direction. The line segment below connects points A and B. A pair of letters with a bar over them like $\overline{AB}$ indicates a line segment.

A   B
•————————————————•

A **line** extends infinitely in each direction. A line is infinite, whereas a line segment is a finite portion of a line. A double arrow placed over a pair of letters like $\overleftrightarrow{AB}$ indicates a line. The arrows below indicate that the line continues forever in each direction.

$$\xleftrightarrow{\quad\quad\quad\overset{\textstyle A}{\bullet}\quad\quad\quad\quad\quad\quad\overset{\textstyle B}{\bullet}\quad\quad\quad}$$

In geometry, the word **line** refers to an infinite length, whereas the term **line segment** refers to the finite distance between two endpoints. In notation, $\overleftrightarrow{AB}$ represents a line passing through points A and B, whereas $\overline{AB}$ represents the line segment connecting points A and B. Note that $\overleftrightarrow{AB}$ is infinite, whereas $\overline{AB}$ is finite.

A **ray** extends infinitely in a single direction. A ray is semi-infinite. A one-sided arrow placed over a pair of letters like $\overrightarrow{AB}$ or $\overleftarrow{AB}$ indicates a ray. In $\overrightarrow{AB}$, the ray begins at A, passes through B, and extends infinitely beyond B (top diagram below). In $\overleftarrow{AB}$, the ray begins at B, passes through A, and extends infinitely beyond A (bottom diagram).

$$\overset{\textstyle A}{\bullet}\xrightarrow{\quad\quad\quad\quad\quad\quad\overset{\textstyle B}{\bullet}\quad\quad\quad}$$
$$\xleftarrow{\quad\quad\quad\overset{\textstyle A}{\bullet}\quad\quad\quad\quad\quad\quad\overset{\textstyle B}{\bullet}\quad\quad\quad}$$

A **horizontal** line runs across to the left and right. A **vertical** line runs up and down. A horizontal line segment (left) and vertical line segment (right) are shown below.

A **plane** extends infinitely in two different directions (such as horizontal and vertical). If a rectangle expanded so much that it became infinite in width and height, it would fill the entire plane. A plane has two dimensions because it extends in two independent directions. This book is focused on plane figures, which are figures that lie within one plane. The **space** of the universe appears to have three dimensions, as it is possible to move along three independent directions: forward and backward, right and left, and up and down.

An **angle** is formed when the endpoints of two line segments join together. The point where the line segments join is called a **vertex**. (Note that "vertex" is singular, whereas "vertices" is plural.) The line segments are called **sides**.

The angle shown above may be represented as either ∠ABC or ∠CBA. (Note that ∠BAC, ∠BCA, ∠CAB, and ∠CBA are different. If three letters are used to represent the angle shown above, point B must be in the middle.) Using three letters to indicate an angle is convenient when a diagram labels points with letters. Sometimes, it is convenient to instead number angles, like ∠3, or to use a lowercase Greek letter, like θ (theta). When using a Greek letter, the angle symbol ∠ will not be used.

Lines, line segments, or rays **intersect** if they cross paths. When two lines intersect, four angles are formed, as shown below.

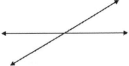

Angles may be measured in degrees or radians. Almost all protractors are marked in degrees. If a circle is divided into 360 equal parts, the angle of each part as measured from the center of the circle (as shown below) equals one **degree**. The degree symbol is a small raised circle (°). For example, 360° represents 360 degrees.

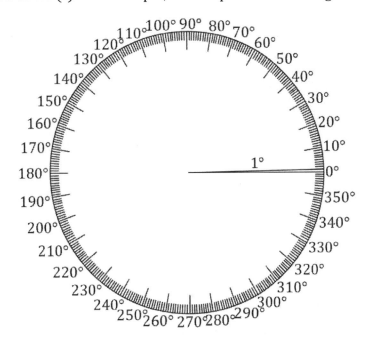

A **radian** is a common alternative to working with degrees. Radians are related to the number π (pi), which approximately equals 3.14159 (though the digits continue on forever without repeating). For any circle, the ratio of its circumference to its diameter is equal to π. Since diameter is twice the radius, the formula for circumference may be expressed as $C = 2\pi r$. The idea behind radians is to let 2π radians (which appears in the circumference formula) correspond to 360°, such that π radians corresponds to 180°. To convert an angle from degrees to radians, multiply by π and divide by 180°. It is not necessary to plug in a numerical value for π. Just leave the symbol in the answer. For example, 60° corresponds to $\frac{60\pi}{180} = \frac{\pi}{3}$ rad. (Note that "rad" is the abbreviation for radians.) To convert an angle from radians to degrees, multiply by 180° and divide by π. For example, $\frac{\pi}{4}$ rad corresponds to $\frac{\pi}{4}\frac{180}{\pi} = \frac{180°}{4} = 45°$.

Why bother with radians? Why not just work with degrees? One reason is that the common form of the arc length formula $s = r\theta$ (Volume 2) only works if the angle is expressed in radians. Many of the problems in this book use degrees because these are familiar, but some problems use radians because it is important to get acquainted with radians (since certain formulas, like the arc length formula, require radians).

A **right angle** measures 90° (which is equivalent to $\frac{\pi}{2}$ rad), corresponding to one-fourth of a circle (as measured from the center). Lines (or line segments) are **perpendicular** if they intersect at right angles. A right angle is often indicated in a diagram by drawing a small square where the lines (or line segments or rays) meet, as shown below. The symbol $\perp$ stands for "perpendicular." For example, $\overline{AB} \perp \overline{BC}$ in the diagram below. Note that angle $\angle ABC$ is a right angle: $\angle ABC = 90°$. For problems that draw horizontal and vertical lines, it helps to remember that horizontal and vertical lines are perpendicular.

A seemingly peculiar, yet important, angle is the **straight** "angle." When three points lie on the same line, they form a straight angle. A straight angle has a measure of 180°

(which is equivalent to π rad). For example, points A, B, and C form a straight angle in the diagram below. Note that angle ∠ABC is a straight angle: ∠ABC = 180°. (Compare with the similar diagram for supplementary angles below to see why it equals 180°.)

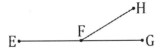

Two angles are **complements** if they add up to 90° (which is equivalent to $\frac{\pi}{2}$ rad). Such angles are called **complementary angles**. When complementary angles join together, they combine to form a right angle, as shown below: ∠ABD + ∠DBC = 90°.

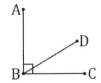

Two angles are **supplements** if they add up to 180° (which is equivalent to π rad). Such angles are called **supplementary angles**. When supplementary angles join together, they combine to form a straight angle, as shown below: ∠EFH + ∠HFG = 180°.

Tip: In the alphabet, C comes before S. Numerically, 90° is less than 180°. This can be used to help remember that complementary angles add up to 90° and supplementary angles add up to 180°. In addition, "supplementary" and "straight angle" each begin with the letter S.

Adjacent angles are beside one another. For example, in the diagrams above, ∠ABD and ∠DBC above are adjacent angles, and ∠EFH and ∠HFG are adjacent angles.

If angles form a **full circle**, the angles add up to 360° (which is equivalent to 2π rad). For example, ∠1 + ∠2 + ∠3 = 360° and α + β + γ + δ = 360° in the diagrams on the next page. Recall that there are three common conventions for labeling angles: three letters like ∠ABD above, a number like ∠1, or a Greek letter like α (alpha), β (beta), γ (gamma), or δ (delta). Since all three conventions are in common usage, it is helpful to be familiar with each of them. This book uses all three common conventions to offer practice working with each method. (Note that a table of the names of the Greek letters can be found at the back of this book.)

Vertical angles appear on opposite sides of a vertex where two lines intersect. When two lines intersect, this creates two pairs of **vertical angles**. Vertical angles are equal. For example, $\alpha = \gamma$ and $\beta = \delta$ in the right diagram above.

An **acute angle** is smaller than 90° (left diagram), an **obtuse angle** is greater than 90° (middle diagram), and a **right angle** equals 90° (right diagram). As noted previously, a small square is often used to indicate when lines (or segments) are perpendicular.

A **reflex angle** refers to an angle that is greater than 180°. For example, angle β (beta) below, which is indicated by the dashed arc, is a reflex angle, whereas angle α (alpha) is acute. Note that $\alpha + \beta = 360°$ because together they form a full circle.

A pair of uppercase letters without a line or arrow over them indicates a **distance** or **length**, like AB. Line segment $\overline{AB}$ has length AB. The notation $\overline{AB}$ represents the line segment connecting A to B, whereas AB represents the distance between them. These are related, but not exactly the same: $\overline{AB}$ refers to the geometric figure, whereas AB is a measured distance. Note that AB is the length of $\overline{AB}$. When a length appears in a formula (like the formula for area, perimeter, or the Pythagorean theorem), then it is common to use a single letter (like L, b, or h) to represent length.

If two different line segments have the same length, the line segments are said to be **congruent**. The **congruence** of two geometric figures is a way of stating that the figures are identical (except that one figure may appear rotated relative to the other). When two line segments are congruent, they have equal lengths. The symbol $\cong$ represents congruence, whereas the symbol $=$ represents equality. The congruence symbol ($\cong$) is used to express that two geometric figures are identical, whereas the equal sign ($=$) is used to express that two measurements are the same. For example, compare $\overline{AB} \cong$

$\overline{CD}$, which states that line segment $\overline{AB}$ is congruent to line segment $\overline{CD}$, with AB = CD, which states that AB and CD are the same length. The two relations go hand-in-hand in the sense that if $\overline{AB}$ is congruent with $\overline{CD}$, it follows that AB and CD are the same length. It is customary to use the symbol ≅ when stating that geometric figures appear identical and the symbol = when stating that two measurements are equal. Note that two angles may be congruent (for which it follows that the measurements of the angles are equal) and that two polygons (such as triangles) may be congruent.

In a diagram, congruent line segments are indicated by drawing a tick mark (|) along each segment. For example, the tick marks in $\overline{AB}$ and $\overline{BC}$ below indicate that $\overline{AB}$ ≅ $\overline{BC}$. It follows that AB = BC (since the line segments are congruent, their lengths are equal).

If there are two pairs of congruent line segments, one pair can have a single tick mark drawn along each segment and the other pair can have double tick marks drawn along each segment. For example, the single tick marks below indicate that $\overline{AB}$ ≅ $\overline{DE}$ while the double tick marks below indicate that $\overline{BC}$ ≅ $\overline{CD}$.

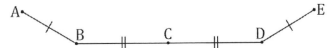

Congruent angles are indicated by drawing small arcs in the angles. For example, arcs indicate that ∠1 ≅ ∠2 in the diagram below. It follows that these angles have equal angular measures (since the angles are congruent, their measures are equal).

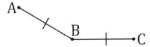

If there are two pairs of congruent angles, single arcs and double arcs can be used to indicate the congruence of each pair. For example, the single arcs below indicate that ∠CBD ≅ ∠EBA while the double arcs below indicate that ∠ABC ≅ ∠DBE.

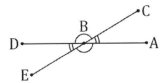

If a line segment or angle is **bisected**, this means that it is cut into two congruent parts. The **midpoint** of a line segment bisects the line segment. Point B is the midpoint in the left diagram, such that $\overline{AB} \cong \overline{BC}$, which means that AB = BC. Line segment $\overline{EG}$ in the right diagram is an angle bisector, such that $\angle DGE \cong \angle EGF$.

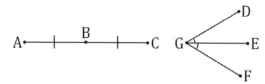

The term **equidistant** means equally distant. For example, point B is equidistant from points A and C in the diagram above (but not in the diagram below).

Three or more points are **collinear** if they lie on the same line. For example, points A, B, and C below on the left are collinear. Three or more lines are **concurrent** if they intersect at the same point. For example, lines $\overleftrightarrow{DG}$, $\overleftrightarrow{EH}$, and $\overleftrightarrow{FI}$ below on the right are concurrent. Two or more geometric figures are **coplanar** if they lie in the same plane. Two geometric figures are **coincident** if they are one and the same.

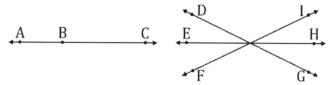

Any two points (that are not coincident) define a line. Given any pair of points (that are not coincident), a single line can be drawn such that the line passes through both points). Any three points that are not collinear define a plane. Given any three points that are not all collinear, a single plane exists that contains all three points.

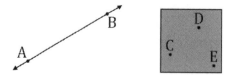

Lines are **parallel** if they extend in the same direction and are the same distance apart at any position such that they will never intersect. If two lines are parallel, they do not intersect and there exists exactly one plane containing both lines. The symbol ∥ is used to indicate that two lines are parallel. For example, $\overleftrightarrow{AB}$ ∥ $\overleftrightarrow{CD}$ in the diagram on the next page. (When writing by hand, ∥ sometimes appears slanted like ∥.)

If two lines **intersect**, they intersect at a single point and there exists exactly one plane containing both lines. If only a portion of each line is drawn, it may be necessary to imagine extending the lines in one direction in order to visualize where they intersect. If two lines lie in the same plane and are not parallel, the lines intersect. The symbol ∦ is used to indicate that two lines are not parallel.

When two **perpendicular** lines intersect, they meet at right angles. Recall that ⊥ is the symbol for perpendicular.

If two lines are not parallel and do not intersect, they are **askew**. Two lines that are askew do not lie within the same plane. (However, it is possible to draw line segments that are askew and lie in the same plane. Recall that a line segment is finite, whereas a line is infinite.)

A **transversal** is a line that intersects at least two other lines. For example, line $\overleftrightarrow{BC}$ is a transversal in the diagram below. When all three lines lie within the same plane, the transversal creates four **interior angles** (∠3, ∠4, ∠5, and ∠6 below) and four **exterior angles** (∠1, ∠2, ∠7, and ∠8 below). There are four pairs of **corresponding** angles in the diagram below: ∠1 and ∠5 correspond, ∠2 and ∠6 correspond, ∠3 and ∠7 correspond, and ∠4 and ∠8 correspond. If the top and bottom lines are parallel $\left(\overleftrightarrow{AB} \parallel \overleftrightarrow{CD}\right)$, then pairs of corresponding angles are congruent (for example, ∠1 ≅ ∠5). There are two pairs of **alternate interior angles**: ∠3 and ∠6, and ∠4 and ∠5. If $\overleftrightarrow{AB} \parallel \overleftrightarrow{CD}$, then alternate interior angles are congruent: ∠3 ≅ ∠6 and ∠4 ≅ ∠5. There are two pairs of **alternate exterior angles**: ∠1 and ∠8, and ∠2 and ∠7. If $\overleftrightarrow{AB} \parallel \overleftrightarrow{CD}$, then alternate exterior angles are congruent: ∠1 ≅ ∠8 and ∠2 ≅ ∠7.

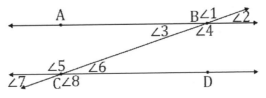

In the previous diagram, any pair of **adjacent angles** have a sum of 180°. For example, $\angle 1 + \angle 2 = 180°$ and $\angle 1 + \angle 3 = 180°$.

If the two lines are parallel $\left(\overleftrightarrow{AB} \parallel \overleftrightarrow{CD}\right)$ in the previous diagram, **same-side interior angles** add up to 180° ($\angle 3 + \angle 5 = 180°$ and $\angle 4 + \angle 6 = 180°$), and **same-side exterior angles** also add up to 180° ($\angle 1 + \angle 7 = 180°$ and $\angle 2 + \angle 8 = 180°$).

According to the **parallel postulate** (which dates back to Euclid), in the previous figure, if the same-side interior angles (either $\angle 3 + \angle 5$ or $\angle 4 + \angle 6$) do not add up to exactly 180°, then lines $\overleftrightarrow{AB}$ and $\overleftrightarrow{CD}$ intersect (and therefore are not parallel). For the case that $\angle 3 + \angle 5 < 180°$, it follows that $\angle 4 + \angle 6 > 180°$ and that lines $\overleftrightarrow{AB}$ and $\overleftrightarrow{CD}$ intersect to the left. For the case that $\angle 3 + \angle 5 > 180°$, it follows that $\angle 4 + \angle 6 < 180°$ and that lines $\overleftrightarrow{AB}$ and $\overleftrightarrow{CD}$ intersect to the right. If instead the same-side interior angles (either $\angle 3 + \angle 5$ or $\angle 4 + \angle 6$) do add up to exactly 180°, then lines $\overleftrightarrow{AB}$ and $\overleftrightarrow{CD}$ must be parallel (in this case, $\overleftrightarrow{AB}$ and $\overleftrightarrow{CD}$ do not intersect). The previous notes regarding the congruence of corresponding angles, the congruence of alternate interior angles, and the congruence of alternate exterior angles only apply if lines $\overleftrightarrow{AB}$ and $\overleftrightarrow{CD}$ are parallel. Similarly, the previous notes regarding same-side interior angles or same-side exterior angles adding up to 180° only apply if lines $\overleftrightarrow{AB}$ and $\overleftrightarrow{CD}$ are parallel. However, the notes regarding adjacent angles (for which there are numerous pairs) apply regardless of whether or not lines $\overleftrightarrow{AB}$ and $\overleftrightarrow{CD}$ are parallel.

An **axiom** or **postulate** refers to a basic principle that is considered to be self-evident. A **theorem** is a mathematical statement that can be proven using postulates (or other theorems that are already known to be true) by applying logical reasoning. A proof applies logical reasoning, given information, definitions, postulates, diagrams, and theorems already known to be true in order to draw a conclusion about a theorem. A **corollary** refers to a mathematical statement that can be obtained from a theorem with very little effort. A **lemma** is a theorem which, once it is proven, helps to prove a more significant theorem; a lemma is referred to as a "helping theorem."

If a mathematical statement has the form, "If A is true, then B is true," then the **converse** of this statement is, "If B is true, then A is true," the **inverse** of the original statement is, "If A is not true, then B is not true," and the **contrapositive** of the original statement is, "If B is not true, then A is not true." Beware that if the original statement is true, the converse and inverse may be false. The converse and inverse, if true, must be proven in addition to the original statement before they can be applied. However, the contrapositive is true if the original statement is true.

Distances, angles, and areas obey the usual rules of arithmetic. For example, if the line segment $\overline{AB}$ is divided up into two shorter line segments, $\overline{AC}$ and $\overline{BC}$ (where C lies on $\overline{AB}$), then the lengths of the shorter line segments add up to the length of the original line segment: $AC + BC = AB$.

The **transitive rule** states that if figure 1 is congruent with figure 2 and if figure 1 is also congruent with figure 3, then it follows that figure 2 is congruent with figure 3. The **reflexive property** states that two figures that are completely coincident are congruent.

Chapter 1 Examples

Example 1. In the diagram below, points D, B, and E are collinear and $\angle DBC = 28°$. Find $\angle ABD$, $\angle CBE$, and $\angle ABE$.

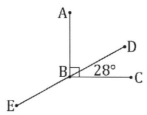

Note that $\angle ABC = 90°$. This means that $\angle ABD$ and $\angle DBC$ are complementary angles.

$$\angle ABD + \angle DBC = 90°$$

$$\angle ABD + 28° = 90°$$

$$\angle ABD = 90° - 28° = 62°$$

Note that $\angle DBC$ and $\angle CBE$ are supplementary angles.

$$\angle DBC + \angle CBE = 180°$$

$$28° + \angle CBE = 180°$$

$$\angle CBE = 180° - 28° = 152°$$

The four angles $\angle ABD$, $\angle DBC$, $\angle CBE$, and $\angle ABE$ form a full circle.

$$\angle ABD + \angle DBC + \angle CBE + \angle ABE = 360°$$

$$62° + 28° + 152° + \angle ABE = 360°$$

$$\angle ABE = 360° - 62° - 28° - 152° = 118°$$

Notes:
- This is not the only way to solve this problem. When possible, it would be wise to solve a problem two different ways and see if the answers are consistent.
- A different way to find $\angle ABE$ is to note that $\angle ABD$ and $\angle ABE$ are supplementary angles: $\angle ABE = 180° - \angle ABD = 180° - 62° = 118°$.
- The answers were determined by applying geometry concepts and performing arithmetic. The values were not found by making measurements.
- Many geometry problems do not draw problems to scale, which means that the problems must be found by applying principles and performing arithmetic, not by making measurements.

Example 2. Determine angle θ (theta) in the diagram below. Also, for all four of the angles in this diagram (including the top and bottom angles), indicate whether each angle is acute, right, or obtuse.

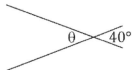

Note that θ and 40° are vertical angles. Since vertical angles are equal, θ = 40°. (The lengths of the line segments have no impact on the angle. Only the directions of the two sides determine the angle. This concept can easily be verified using a protractor.) The left and right angles are acute. The top and bottom angles are obtuse. (Since the top angle and right angle are supplementary, it is possible to determine that the top angle equals 180° − 40° = 140°, and similarly for the bottom angle.)

Example 3. In the diagram below, $\angle ABC = \frac{5\pi}{6}$ rad. Determine ∠1, ∠2, and ∠3 in degrees.

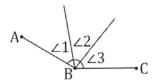

The small single arcs indicate that ∠1, ∠2, and ∠3 are congruent.

$$\angle 1 + \angle 2 + \angle 3 = 3(\angle 1) = \frac{5\pi}{6} \text{ rad}$$

$$\angle 1 = \frac{5\pi}{18} \text{ rad}$$

To convert from radians to degrees, multiply by 180° and divide by π.

$$\angle 1 = \frac{5\pi}{18} \times \frac{180°}{\pi} = 50°$$

Notes:

- ∠1 = ∠2 = ∠3 = 50° because these three angles are congruent.

- Check the answer: $\angle 1 + \angle 2 + \angle 3 = 50° + 50° + 50° = 150° = \frac{150\pi}{180} = \frac{5\pi}{6}$ rad.

- Divide by 3 on both sides of $3(\angle 1) = \frac{5\pi}{6}$ to get $\angle 1 = \frac{5\pi}{18}$. (In conventional algebra notation, these are like $3x = \frac{5\pi}{6}$ and $x = \frac{5\pi}{18}$.)

Example 4. In the diagram below, AE is 6 units long. What is AB?

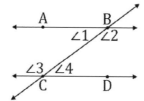

The single tick marks indicate that $\overline{AB}$, $\overline{BC}$, $\overline{CD}$, and $\overline{DE}$ are congruent line segments. Their lengths are equal: AB = BC = CD = DE. These distances add up to AE.

$$AB + BC + CD + DE = AE$$

$$4AB = 6$$

$$AB = \frac{6}{4} = \frac{3}{2} = 1.5$$

The distance AB equals 1.5 units (which is equivalent to the fraction 3/2).

Example 5. In the diagram below, $\overleftrightarrow{AB} \parallel \overleftrightarrow{CD}$ and $\angle 1 = 36°$. Find $\angle 3$ and $\angle 4$.

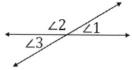

Same-side interior angles add up to 180°. Angles $\angle 1$ and $\angle 3$ are same-side interior angles.

$$\angle 1 + \angle 3 = 180°$$

$$\angle 3 = 180° - \angle 1 = 180° - 36° = 144°$$

Alternate interior angles are congruent: $\angle 4 = \angle 1 = 36°$.

Example 6. Prove that vertical angles are congruent.

Vertical angles form when two lines intersect. The diagram above shows two lines that intersect. Since $\angle 1$ and $\angle 2$ form a straight angle, they must be supplementary angles: $\angle 1 + \angle 2 = 180°$. Subtract $\angle 1$ from both sides: $\angle 2 = 180° - \angle 1$. Since $\angle 2$ and $\angle 3$ also form a straight angle, they must be supplementary angles: $\angle 2 + \angle 3 = 180°$. Replace $\angle 2$ with $180° - \angle 1$ in the equation $\angle 2 + \angle 3 = 180°$ to get $180° - \angle 1 + \angle 3 = 180°$. Subtract 180° from both sides. Note that 180° cancels out: $-\angle 1 + \angle 3 = 0$. Add $\angle 1$ to both sides: $\angle 3 = \angle 1$. Note that $\angle 1$ and $\angle 3$ are vertical angles.

Example 7. Consider the statement, "If two lines are parallel, they do not intersect."

(A) Is this statement true?

Yes. Parallel lines, by definition, are the same distance apart at any position such that they never intersect.

(B) What is the converse of the given statement? Is the converse necessarily true?

The converse of the statement is, "If two lines do not intersect, the lines are parallel." The converse is not true in general. If two lines do not intersect, one possibility is that the lines are parallel, but another possibility is that the lines are askew. (If the original statement had restricted the lines to lie within the same plane, in that case the converse would have been true. If two lines that lie within the same plane do not intersect, the lines are parallel. However, the original statement did not impose any such restriction.)

(C) What is the inverse of the given statement? Is the inverse necessarily true?

The inverse of the statement is (be sure to look at the original statement and not the converse from Part B), "If two lines are not parallel, they intersect." The inverse is not true in general. If two lines are not parallel, one possibility is that the lines intersect, but another possibility is that the lines are askew.

(D) What is the contrapositive of the given statement? Is the contrapositive true?

The contrapositive of the statement is (be sure to look at the original statement and not the answers to Parts B or C), "If two lines intersect, the lines are not parallel." This is certainly true. (In fact, the contrapositive is always true if the original statement is true.)

Chapter 1 Problems

Notes: The diagrams are not drawn to scale. Apply principles of geometry, logic, and reasoning to solve the problems. (Do not make measurements to solve the problems.)

1. In the diagram below, $\angle AEB = 34°$, $\angle AEC = 90°$, and $\angle BED = 90°$. What is $\angle CED$?

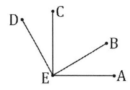

2. Determine $\angle 1$, $\angle 2$, and $\angle 3$ in the diagram below.

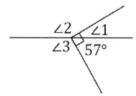

3. In the diagram below, $\frac{\pi}{6}$ is in radians. Find θ, φ, and ψ in radians and in degrees.

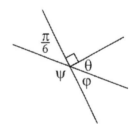

4. In the diagram below, $\angle AED = 90°$. As indicated by the single arcs, three angles are congruent. Determine $\angle AEB$, $\angle BEC$, and $\angle CED$ in degrees and in radians.

5. In the diagram below, points A, B, and C are collinear and five angles are congruent. Determine θ in degrees.

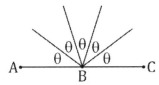

6. In the diagram below, three concurrent lines make six congruent angles. Determine φ in degrees and in radians.

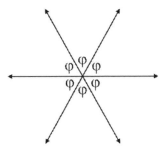

7. In the diagram below, ∠AEB is a right angle, the acute angle is one-third of the right angle, and the two obtuse angles are congruent. Express ∠AEB, ∠BEC, ∠CED, and ∠DEA in degrees and in radians.

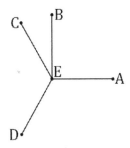

8. In the diagram below, points A, F, and D are collinear, and points B, F, and E are also collinear. Determine ∠1, ∠2, and ∠3 in the diagram below.

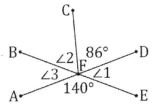

9. In the diagram below, $\frac{2\pi}{3}$ is in radians. Angle θ is a reflex angle. Find θ in radians and in degrees.

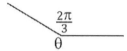

10. In the diagram below, $\overline{AB} \perp \overline{BD}$ and ∠ABC is twice ∠CBD. Find ∠ABC and ∠CBD.

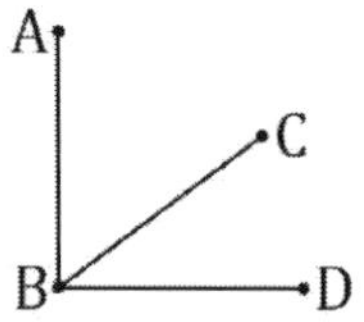

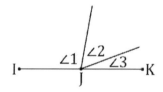

11. In the diagram below, points I, J, and K are collinear. Angles ∠1, ∠2, and ∠3 come in the ratio 5:3:1. Determine ∠1, ∠2, and ∠3 in degrees.

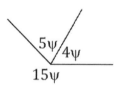

12. Determine ψ in degrees and in radians in the diagram below.

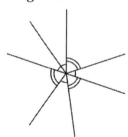

13. In the diagram below, an angle indicated by a double arc is twice an angle indicated by a single arc. Find each angle in degrees.

14. In the diagram below, CD is 2.4 units long. Determine AD.

15. In the diagram below, PR is 9 units long. Determine PU.

16. Points A thru J in the diagram below are collinear and equally spaced. The distance from A to J is 36 units. Find AB, CF, and BH.

17. In the diagram below, GK is 12 units long and HI is twice as long as GH. Determine GH and HI.

18. In the diagram below, VZ is 3 units long, W is the midpoint of $\overline{VZ}$, X is the midpoint of $\overline{WZ}$, and Y is the midpoint of $\overline{XZ}$. Determine XY.

V•————————————W X Y •Z

19. In the diagram below, $\overleftrightarrow{AC}$ ∥ $\overleftrightarrow{FH}$ and ∠ADE = 54°. Find ∠FEG, ∠DEH, and ∠GEH.

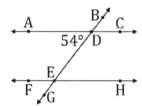

20. In the diagram below, $\overleftrightarrow{AB}$ ∥ $\overleftrightarrow{CD}$. Find ∠1, ∠2, ∠3, and ∠4.

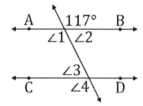

21. For each diagram below, indicate whether or not the left and right lines are parallel. If they are not parallel, indicate whether they intersect above or below the transversal.

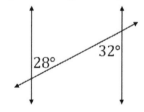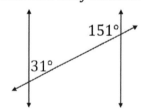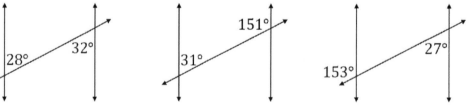

22. In the diagram below, $\overline{TU}$ ∥ $\overline{VW}$ and $\overline{WX}$ ∥ $\overline{YZ}$. Determine ∠VWX.

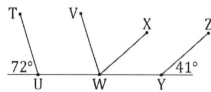

23. In the diagram below, $\overline{AB} \parallel \overline{DE}$. The given angle is in radians. Find θ in radians and in degrees.

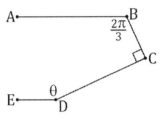

24. In the diagram below, $\overline{KL} \perp \overline{MN}$. Determine $\angle LMN$.

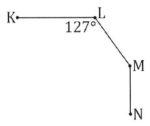

25. In the diagram below, $\overline{QR} \parallel \overline{ST}$ and $\overline{RS} \parallel \overline{TU}$. Find $\angle 1$ and $\angle 2$.

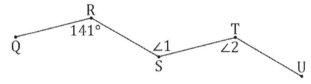

26. If two angles are supplementary angles, how many of the angles can be acute? How many can be right? How many can be obtuse? What determines how many of which type of angles there will be?

27. If three angles add up to 180°, how many of the angles can be acute? How many can be right? How many can be obtuse? What determines how many of which type of angles there will be?

28. If $\overleftrightarrow{AB} \parallel \overleftrightarrow{CD}$ in the diagram below, prove that $\angle 1 \cong \angle 4$. (These are alternate interior angles.)

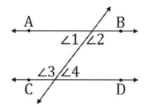

29. If $\overleftrightarrow{AB} \parallel \overleftrightarrow{CD}$ in the diagram below, prove that $\angle 2 \cong \angle 6$. (These are corresponding angles.)

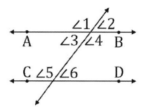

30. Consider the statement, "If two lines are perpendicular, they are not parallel."

(A) Is this statement true?

(B) What is the converse of the given statement? Is the converse true?

(C) What is the inverse of the given statement? Is the inverse true?

(D) What is the contrapositive of the given statement? Is the contrapositive true?

Angles of a Triangle

This chapter explores basic properties of triangles relating to angles and sides. Topics includes angle sum theorems for interior and exterior angles, and different kinds of triangles (acute, right, obtuse, scalene, isosceles, and equilateral).

Chapter 2 Concepts

A **triangle** is formed when three points (that are not all collinear) are joined by three line segments, as illustrated below. Points A, B, and C below are called **vertices**. Line segments $\overline{AB}$, $\overline{BC}$, and $\overline{CA}$ form the **sides** (or edges) of the triangle. The notation for the triangle below is ΔABC. (The order of the letters in ΔABC is not important.)

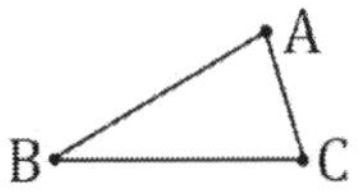

There are two different types of angles that are often of interest regarding a triangle. The three **interior angles** form at the three vertices and lie inside of the triangle. For the triangle above, the interior angles are ∠BAC, ∠ABC, and ∠BCA. An **exterior** angle forms between one side of the triangle and a line that extends from an adjacent side. In the triangle below, α is an interior angle while β is an exterior angle.

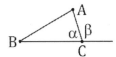

An **isosceles** triangle has two sides with equal length. An **equilateral** triangle has three sides with equal length. A **scalene** triangle does not have any two sides of equal length.

The triangle below on the left is isosceles, the middle triangle below is equilateral, and the triangle below on the right is scalene.

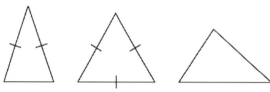

An **equiangular** triangle has three equal interior angles. A triangle that is equilateral is also equiangular, and vice-versa. (However, as will be discussed in Chapters 9-10, shapes with four or more sides can be equilateral without being equiangular, and vice-versa.) An **equilateral** triangle has three 60° interior angles (which are equivalent to $\frac{\pi}{3}$ rad).

In an **isosceles** triangle, the two interior angles that are opposite to the congruent sides are congruent.

The sum of the three **interior angles** of any triangle equals 180° (which is equivalent to π rad). For example, $\angle 1 + \angle 2 + \angle 3 = 180°$ for the triangle below.

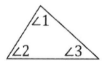

The sum of any two **interior angles** for any triangle equals the **exterior angle** for the remaining vertex. For example, $\angle 1 + \angle 2 = \angle 4$ for the triangle below.

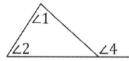

At any vertex of any triangle, there are two possible exterior angles; these two possible **exterior angles** are equal. For example, $\angle 4 = \angle 6$ below. ($\angle 5$ is not an exterior angle.)

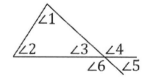

The sum of the three **exterior angles** of any triangle equals 360° (which is equivalent to 2π rad), provided that exactly one exterior angle is included for each vertex. For example, $\angle 4 + \angle 5 + \angle 6 = 360°$ for the triangle below.

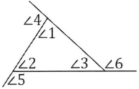

An **acute triangle** has three acute interior angles (left diagram), an **obtuse triangle** has one obtuse interior angle (middle diagram), and a **right triangle** has one right interior angle (right diagram).

Chapter 2 Examples

Example 1. Determine ∠3 in the diagram below.

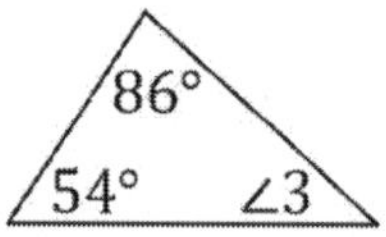

The three interior angles of a triangle add up to 180°.

$$86° + 54° + ∠3 = 180°$$
$$140° + ∠3 = 180°$$
$$∠3 = 180° - 140° = 40°$$

Example 2. Determine θ in the diagram below.

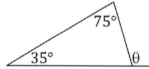

The sum of two interior angles equals the exterior angle for the remaining vertex.

$$35° + 75° = 110° = θ$$

Example 3. Determine ∠BCD in both radians and degrees in the diagram below.

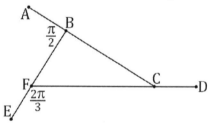

The sum of three exterior angles for a triangle is $2π$ rad (equivalent to 360°) provided that one exterior angle is included for each vertex.

$$∠ABF + ∠BCD + ∠CFE = 2π$$
$$\frac{π}{2} + ∠BCD + \frac{2π}{3} = 2π$$
$$\frac{7π}{6} + ∠BCD = 2π$$
$$∠BCD = 2π - \frac{7π}{6} = \frac{5π}{6} \text{ rad} = 150°$$

Notes:

- To add two fractions, first find a common denominator: $\frac{1}{2} + \frac{2}{3} = \frac{3}{6} + \frac{4}{6} = \frac{7}{6}$.
- To subtract a fraction, find a common denominator: $2 - \frac{7}{6} = \frac{12}{6} - \frac{7}{6} = \frac{5}{6}$.
- In degrees, the three exterior angles are 90°, 150°, and 120°.
- Check the answer using degrees: $90° + 150° + 120° = 360°$.

Example 4. Describe the triangle below.

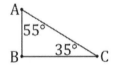

Do not guess if the triangle is acute, right, or obtuse. Solve for the missing angle to find out. The three interior angles add up to 180°.

$$55° + \angle ABC + 35° = 180°$$
$$90° + \angle ABC = 180°$$
$$\angle ABC = 180° - 90° = 90°$$

Since $\angle ABC = 90°$, it is clear that this is a right triangle. More precisely, this is a right scalene triangle. Since all three interior angles are different, the triangle is evidently not isosceles.

Example 5. Prove that the three interior angles of a triangle add up to 180°.

Draw a triangle like the one above. Draw line segment $\overline{AC}$ parallel to the bottom side. This creates two pairs of alternate interior angles: $\angle 2 \cong \angle 4$ and $\angle 3 \cong \angle 5$. Points A, B, and C are collinear such that $\angle ABC$ is a straight angle: $\angle ABC = 180°$. Angles $\angle 4$, $\angle 1$, and $\angle 5$ add up to 180°.

$$\angle 4 + \angle 1 + \angle 5 = 180°$$

Knowing that $\angle 2 \cong \angle 4$ and $\angle 3 \cong \angle 5$ allows $\angle 4$ to be replaced with $\angle 2$ and $\angle 5$ to be replaced with $\angle 3$.

$$\angle 2 + \angle 1 + \angle 3 = 180°$$

Chapter 2 Problems

Notes: The diagrams are not drawn to scale. Apply principles of geometry, logic, and reasoning to solve the problems. (Do not make measurements to solve the problems.)

1. Determine θ in the diagram below.

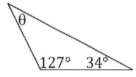

2. Determine ∠BCA in the diagram below.

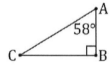

3. Determine ∠2 in both radians and degrees in the diagram below. (The given angles are in radians.)

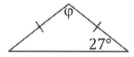

4. Determine φ in the diagram below.

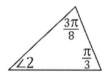

5. Determine ∠1 in both radians and degrees in the diagram below. (The given angle is in radians.)

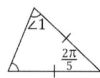

6. In the diagram below, points A, B, and C are collinear. Determine ∠ABD.

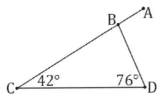

7. Determine α (an exterior angle) in the diagram below.

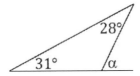

8. Determine ∠4 (an exterior angle) in both radians and degrees in the diagram below. (The given angles are in radians.)

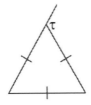

9. Determine τ (an exterior angle) in both radians and degrees in the diagram below.

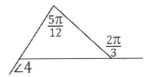

10. Determine ∠BEF (an exterior angle) in the diagram below.

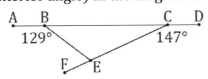

11. Determine ∠1, ∠2, ∠3, ∠4, ∠5, ∠6, ∠7, ∠8, ∠9 and ∠10 in the diagram below.

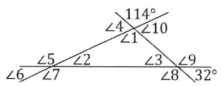

12. In the diagram below, points A, B, and C are collinear and $\overline{AC} \parallel \overline{DE}$. Determine ∠ABD, ∠DBE, and ∠BED. (No angle in this diagram is an exterior angle.)

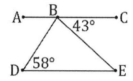

13. Two angles below are marked as right angles. Find θ, φ, and χ.

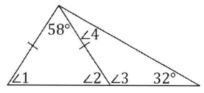

14. Two sides below are marked as congruent. Find ∠1, ∠2, ∠3, and ∠4.

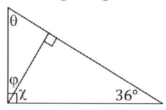

15. Determine α, β, γ, and δ in the diagram below.

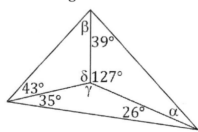

16. For each triangle below, which of the following terms apply: acute, right, obtuse, scalene, isosceles, equilateral, equiangular? (The rightmost angle is given in radians.)

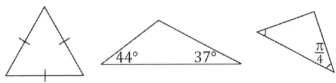

17. Prove that the sum of any two interior angles of a triangle equals the exterior angle for the remaining vertex.

18. At any vertex of any triangle, there are two possible exterior angles. Prove that these two possible exterior angles are congruent.

19. Prove that the sum of the three exterior angles of a triangle equals 360°, provided that exactly one exterior angle is included for each vertex.

20. Prove that an equilateral triangle has 60° interior angles and 120° exterior angles.

21. Consider the statement, "If a triangle is a right triangle, the two smallest interior angles are complementary angles."

(A) Is this statement true?

(B) What is the converse of the given statement? Is the converse true?

(C) What is the inverse of the given statement? Is the inverse true?

(D) What is the contrapositive of the given statement? Is the contrapositive true?

Congruent Triangles

This chapter focuses on triangles that are congruent. There are four ways to test for congruent triangles: three congruent sides (SSS), two congruent sides and a congruent angle between them (SAS), two congruent angles and a congruent side between them (ASA), and two corresponding congruent angles and a corresponding congruent side not between them (AAS). If two triangles are known to be congruent, it is often helpful to apply the principle that corresponding parts of congruent triangles are congruent (CPCTC).

Chapter 3 Concepts

Two triangles are **congruent** if they have the same shape and size. Congruent triangles have three pairs of congruent sides and three pairs of congruent interior angles. They are basically "identical," but the word "congruent" is used rather than "identical." The two triangles below are congruent. The single, double, and triple tick marks indicate that three pairs of sides are congruent, and the single, double, and triple angle marks indicate that the three pairs of corresponding angles are congruent.

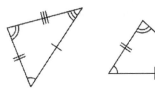

Note that two triangles may be congruent even if one is rotated relative to the other or even if the triangles appear to be mirror images, as shown on the following page.

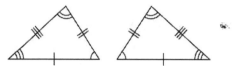

Although three pairs of sides and three pairs of corresponding interior angles are all congruent if two triangles are congruent, it is not necessary to show that all six pairs are congruent in order to know that two triangles are congruent. There are four ways to test for congruence between two triangles.

If all three sides are congruent, it follows that the triangles are congruent. This test is abbreviated **SSS** (for side-side-side). For example, all three sides are congruent in the two triangles below. These triangles are congruent according to SSS.

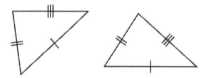

If two sides are congruent and the angle formed by the congruent sides is congruent, it follows that the triangles are congruent. This test is abbreviated **SAS** (for side-angle-side). The congruent angle must be in between the two sides. (Note that "SSA" is not a valid test for congruence. This will be demonstrated later.) For example, two sides and the angle between them are congruent in the triangles below. These triangles are congruent according to SAS.

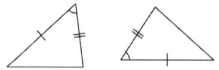

If two angles are congruent and the side that touches both of the congruent angles is congruent, it follows that the triangles are congruent. This test is abbreviated **ASA** (for angle-side-angle). The congruent side touches both congruent angles in this case. For example, two angles and the side between them are congruent in the triangles below. These triangles are congruent according to ASA.

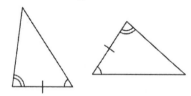

If two angles are congruent and a side that does not touch both of the congruent angles is congruent, if the two angles and side correspond to one another, it follows that the triangles are congruent. This test is abbreviated **AAS** (for angle-angle-side). The angles and side must correspond to one another in the two triangles. For example, two angles and a side that does not touch both angles are congruent in the triangles below, and these angles and side correspond to one another in the two triangles. The way that they correspond is that in both triangles the side with a single tick mark touches the angle with a single arc (but not the angle with the double arc). These triangles are congruent according to AAS.

It is important to check that the angles and side properly correspond before applying AAS. For example, compare the two pairs of triangles below. The two triangles on the left are congruent according to AAS because the congruent side in each case touches the angle with the single arc. In contrast, AAS does not apply to the two triangles on the right because the side that is congruent touches the double arc in one triangle, but the single arc in the other triangle. Study the two pairs below carefully.

AAS   not AAS

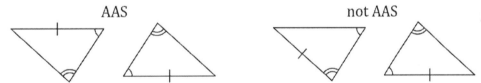

Although SAS is a valid method for establishing congruence, note that "SSA" is not. If the angle that is congruent touches both congruent sides, the triangles are congruent according to SAS. If the angle only touches one of the two congruent sides, this is not sufficient to determine that the triangles are congruent. SAS is valid; "SSA" is not.

SAS is valid   "SSA" is not valid

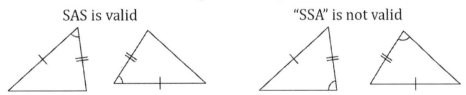

The problem with "SSA" is that if two sides are congruent and an angle that does not touch both of the congruent sides is congruent, it is possible that the two triangles are not congruent, as shown on the following page.

The diagram above illustrates the problem with "SSA." Each triangle has one side that measures 5 units, one side that measures 6 units, and a 25° angle formed by the side that measures 6 units and the third side of unknown length. The two triangles are not congruent. In the triangle on the left, the angle between the 5 and 6 is obtuse, but in the triangle on the right, the angle between the 5 and 6 is acute. The diagram below shows both of these triangles in the same picture. If the side measuring 6 units and the 25° angle are drawn first, there are two different ways that a side measuring 5 units can be drawn, as shown below. Therefore, "SSA" has ambiguity, which is why it is not valid for testing for congruence.

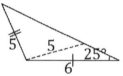

In contrast, SAS is valid for testing congruence because it does not have any ambiguity. For example, if sides measuring 5 units and 6 units are drawn with a 25° angle between them, this results in a single possible triangle, as shown below.

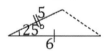

Once it is established that two triangles are congruent, it follows that corresponding parts of congruent triangles are congruent. This is abbreviated CPCTC. It is common for proofs in geometry to first apply either SSS, SAS, ASA, or AAS to establish that two triangles are congruent, and in the next step apply the CPCTC to establish that two sides or angles are congruent.

Chapter 3 Examples

Example 1. Prove that the top and bottom triangles below are congruent.

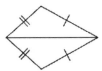

The single and double tick marks indicate that two pairs of sides are congruent. The third side is also congruent (through the reflexive property) because it is shared by both of the triangles. Since all three sides are congruent, the triangles are congruent according to SSS.

Example 2. In the diagram below, A, C, and D are collinear. Prove that $\triangle ABD \cong \triangle CBD$.

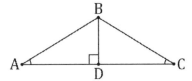

Single arcs indicate that $\angle BAD \cong \angle BCD$. The small square indicates that $\angle BDA = 90°$. Since $\angle BDA$ and $\angle BDC$ are supplementary angles, it follows that $\angle BDC = 90°$. Thus, $\angle BDA \cong \angle BDC$. (They are both right angles.) Both triangles share side BD. Since two angles are congruent and the side that touches the 90° angle is also congruent, the triangles are congruent according to AAS.

Example 3. The two triangles below are congruent. Label the missing side and angles for the left triangle. Note: a, b, and c are the lengths of the sides.

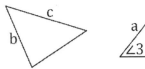

Corresponding parts of congruent triangles are congruent (CPCTC). In each triangle, $\angle 3$ lies between a and c, $\angle 2$ lies between b and c, and $\angle 1$ lies between a and b.

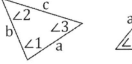

Chapter 3 Problems

Note: The diagrams are not drawn to scale.

1. In the diagram below, $\overline{AB} \parallel \overline{DE}$ and C lies on $\overline{AE}$ and $\overline{BD}$. Prove that $\triangle ABC \cong \triangle CDE$.

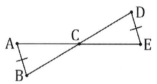

2. In the diagram below, H is the midpoint of $\overline{FI}$ and H is also the midpoint of $\overline{GJ}$. Prove that $\triangle FGH \cong \triangle HIJ$.

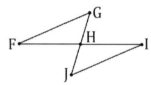

3. Prove that the top and bottom triangles below are congruent.

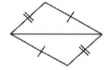

4. In the diagram below, K, M, and N are collinear and L, M, and O are collinear. Prove that $\triangle KLM \cong \triangle MNO$.

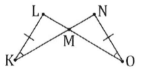

5. Prove that $\triangle PRS \cong \triangle QRS$ in the diagram below.

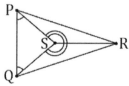

6. In the diagram below, Z is the midpoint of $\overline{WY}$. Prove that ΔWXZ ≅ ΔXYZ.

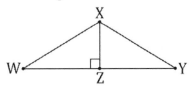

7. Prove that ΔABE ≅ ΔBCD in the diagram below.

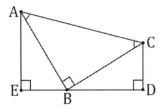

8. Prove that ΔPST ≅ ΔQRT in the diagram below.

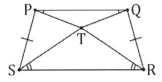

9. In the diagram below, $\overline{GJ} \perp \overline{FG}$ and $\overline{GJ} \perp \overline{IJ}$. Prove that ΔFGH ≅ ΔHIJ.

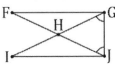

10. In the diagram below, K, M, and N are collinear. Prove that ΔLMN ≅ ΔKNO.

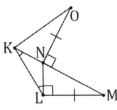

11. In the diagram below, $\overline{AC} \parallel \overline{BD}$, $\overline{AD} \parallel \overline{BE}$, and D is the midpoint of $\overline{CE}$. Prove that ΔACD ≅ ΔBDE.

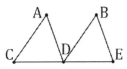

12. The two triangles below are congruent. Label the missing sides and angle for the left triangle. Note: a, b, and c are the lengths of the sides.

13. The two triangles below are congruent. Label the missing sides and angles for the left triangle. Note: j, q, and x are the lengths of the sides.

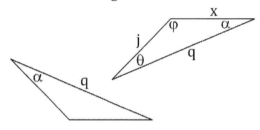

14. Consider the statement, "If two triangles are congruent, all three interior angles are congruent."

(A) Is this statement true?

(B) What is the converse of the given statement? Is the converse true?

(C) What is the inverse of the given statement? Is the inverse true?

(D) What is the contrapositive of the given statement? Is the contrapositive true?

Similar Triangles

This chapter focuses on triangles that are similar, but which are not congruent. There are three ways to test for similar triangles: two congruent angles (AA), two sides that come in the same proportion and a congruent angle between them (SAS similarity), and three sides that come in the same proportion (SSS similarity). If two triangles are known to be similar, it is often helpful to apply the principle that corresponding angles are congruent or the principle that corresponding sides make the same ratios.

Chapter 4 Concepts

Two triangles are **similar** if they have the same shape, but not the same size. Similar triangles have three pairs of congruent angles. The two triangles below are similar. The single, double, and triple angle marks indicate that the three pairs of corresponding angles are congruent. The two triangles below are similar, but they are not congruent.

Note that two triangles may be similar even if one is rotated relative to the other, as shown below, or even if one is reflected relative to the other.

There are three ways to test if two triangles are similar. These tests resemble the tests for congruence.

If two interior angles are congruent, it follows that the triangles are similar. This test is abbreviated **AA** (for angle-angle). Since the three interior angles of a triangle add up to 180°, if two interior angles are congruent, the third interior angle must also be congruent. This is why this test is abbreviated AA (rather than AAA); once two angles are known to be congruent, the third is automatically congruent. For example, the two triangles below are similar according to AA since two angles are marked as congruent.

If two sides make the same ratio and the angle between the sides is congruent, it follows that the triangles are similar. This test is abbreviated **SAS similarity** (for side-angle-side). The word "similarity" at the end distinguishes the SAS similarity test from the SAS test for congruence. As an example, consider the two triangles below. Since the ratio 4/6 equals the ratio 6/9 (since each ratio reduces to 2/3) and since the angle between these sides is marked as congruent, the triangles are similar according to SAS similarity.

If the ratio of all three sides is the same, it follows that the triangles are similar. This test is abbreviated **SSS similarity** (for side-side-side). The word "similarity" at the end distinguishes the SSS similarity test from the SSS test for congruence. As an example, consider the two triangles below. Since the ratio 8:10:12 equals the ratio 4:5:6, the triangles are similar according to SSS similarity.

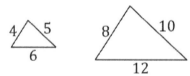

Corresponding sides of similar triangles come in the same proportions. Corresponding angles of similar triangles are congruent. Once two triangles are known to be similar, these two principles may be applied in order to determine unknown sides and angles. For example, consider the two triangles on the following page. Since the two triangles are similar, it is possible to solve for the unknown side labeled x by setting up a proportion. It is important to make sure that the edge lengths correspond in the ratios.

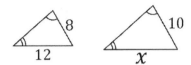

In the diagram above, 8 and 10 are corresponding sides (each is opposite to an angle with a double arc) and 12 and x are corresponding sides (each is opposite to an angle with a single arc). In similar triangles, corresponding sides make the same ratio. For the similar triangles above, this means that the ratio 8/12 equals the ratio 10/x. Note that the 8 and 12 are sides of the triangle on the left and the 10 and x are corresponding sides of the triangle on the right. Set these ratios equal to one another.

$$\frac{8}{12} = \frac{10}{x}$$

Cross multiply. This is equivalent to multiplying both sides of the equation by both denominators. That is, multiply both sides by $12x$.

$$8x = 120$$

Divide both sides of the equation by 8.

$$x = \frac{120}{8} = 15$$

The answer is $x = 15$. Check the answer: 8/12 and 10/15 both reduce to 2/3.

The two triangles below are similar. Since the corresponding sides of similar triangles come in the same proportion, this means that AB:BC:CA = DE:EF:FD. This proportion can be expressed using fractions three ways: $\frac{AB}{BC} = \frac{DE}{EF}$, $\frac{BC}{CA} = \frac{EF}{FD}$, and $\frac{AB}{CA} = \frac{DE}{FD}$. Each of these fractions involves one pair of corresponding sides.

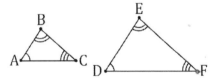

The symbol ~ is used to indicate similarity. For example, $\triangle ABC \sim \triangle DEF$ represents that triangle ABC is similar to triangle DEF. As usual, the order of the letters in $\triangle ABC$ does not matter (whereas the order of the letters in $\angle ABC$ does matter). An expression like $\triangle ABC \sim \triangle DEF$ does not indicate which vertices correspond to one another. Instead, it is necessary to study the diagrams to determine which angles and sides correspond. In the diagram above, AB corresponds to DE and BC corresponds to EF.

Chapter 4 Examples

Example 1. In the diagram below, $\overline{BE} \parallel \overline{CD}$. Prove that $\triangle ABE \sim \triangle ACD$. Find DE.

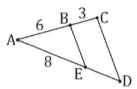

Note that $\angle BAE \cong \angle CAD$ because this angle is shared by both triangles. Also note that $\angle AEB \cong \angle ADC$ since these are corresponding angles (since $\overline{BE} \parallel \overline{CD}$). Since two angles are congruent, the triangles are similar according to AA.

Since the triangles are similar, corresponding sides make the same ratio. According to the diagram, $AB = 6$, $BC = 3$, and $AE = 8$. Observe that $AC = AB + BC = 6 + 3 = 9$. Note that AB and AE of $\triangle ABE$ correspond to AC and AD of $\triangle ACD$. The ratio AB/AE in $\triangle ABE$ equals the ratio AC/AD in $\triangle ACD$.

$$\frac{AB}{AE} = \frac{AC}{AD}$$

$$\frac{6}{8} = \frac{9}{AD}$$

Cross multiply (or multiply both sides by 8 and multiply both sides by AD).

$$6AD = 8(9)$$

$$6AD = 72$$

$$AD = \frac{72}{6} = 12$$

$$DE = AD - AE = 12 - 8 = 4$$

Check the answers: $AB/AE = 6/8 = 3/4$ agrees with $AC/AD = 9/12 = 3/4$.

Example 2. Prove that $\triangle ACD \sim \triangle ABC \sim \triangle ABD$ in the diagram below. Find $\angle BAD$ and AD.

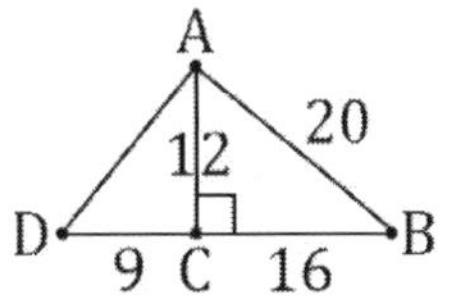

Note that $\angle ACD \cong \angle ACB$ because each is a right angle. The ratio of DC to AC is 9/12 = 3/4 and the ratio of AC to BC is 12/16 = 3/4. Since the right angles are congruent and the sides touching the right angles have the same ratio (3/4), $\triangle ACD \sim \triangle ABC$ according to SAS similarity.

Since $\triangle ACD \sim \triangle ABC$, the corresponding angles must be congruent: $\angle ADC \cong \angle BAC$ and $\angle CAD \cong \angle ABC$. Note that $\angle ADC$ and $\angle ABC$ are two interior angles of $\triangle ABD$. Since two interior angles of $\triangle ABD$ are congruent with two interior angles of $\triangle ACD$ and $\triangle ABC$, all three triangles ($\triangle ACD$, $\triangle ABC$, and $\triangle ABD$) are similar according to AA.

Note that $\angle BAD$ of $\triangle ABD$ corresponds to $\angle ACD$ of $\triangle ACD$ and $\angle ACB$ of $\triangle ABC$. Since the three triangles are similar, it follows that $\angle BAD = 90°$.

Since the triangles are similar, corresponding sides have the same ratio. Note that DC and AD of $\triangle ACD$ correspond to AC and AB of $\triangle ABC$. (These are the shortest and longest sides, respectively.) The ratio DC/AD in $\triangle ABD$ equals the ratio AC/AB in $\triangle ABC$.

$$\frac{DC}{AD} = \frac{AC}{AB}$$

$$\frac{9}{AD} = \frac{12}{20}$$

Cross multiply (or multiply both sides by AD and multiply both sides by 20).

$$9(20) = 12AD$$

$$180 = 12AD$$

$$\frac{180}{12} = 15 = AD$$

Check the answers: DC:AC:AD = 9:12:15 agrees with AC:BC:AB = 12:16:20 and with AD:AB:BD = 15:20:25. All three triple ratios reduce to 3:4:5. (Multiply 3, 4, and 5 each by 3 to make 9, 12, and 15. Multiply 3, 4, and 5 each by 4 to make 12, 16, and 20. Multiply 3, 4, and 5 each by 5 to make 15, 20, and 25.)

Chapter 4 Problems

Note: The diagrams are not drawn to scale.

1. In the diagram below, $\overline{AB} \parallel \overline{DE}$. Prove that $\triangle ABC \sim \triangle CDE$. Find BC.

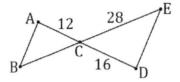

2. Prove that $\triangle VWX \sim \triangle XYZ$ in the diagram below. Find $\angle VWX$ and $\angle WVX$.

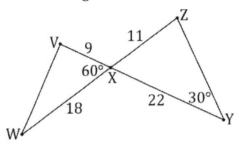

3. For the diagram below, prove that $\triangle FGJ \sim \triangle FHI$ and prove that $\overline{GJ} \parallel \overline{HI}$. Find HI.

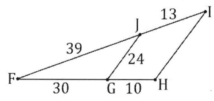

4. Prove that $\triangle RSU \sim \triangle STU$ in the diagram below. Find ST and TU.

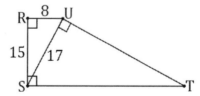

5. Prove that ΔKMN~ΔKLM in the diagram below. Which angles are congruent?

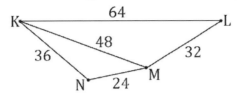

6. In the diagram below, $\overline{AF} \parallel \overline{BC}$ and $\overline{EF} \parallel \overline{CD}$. Prove that ΔAEF~ΔBCD. Find AF and DE.

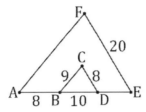

7. Prove that ΔPQS~ΔQRS~ΔPQR in the diagram below. Find QR and RS.

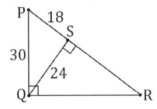

8. Prove that ΔXYZ~ΔVWY in the diagram below. Find VZ.

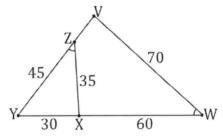

9. In the diagram below, $\overline{TU} \parallel \overline{WV}$ and $\overline{UV} \parallel \overline{WX}$. Prove that ΔTUV~ΔVWX. Find TU and UV. Note that TX = 96 and VX = 24.

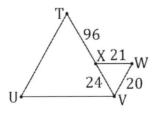

10. In the diagram below, AD = 16, AB = BD, and AC = BC. Prove that ΔABD∼ΔABC. Find BC and CD.

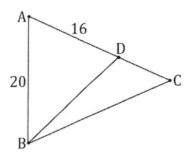

11. In the diagram below, KM = 81 and LN = 100. Which triangles are similar? Which angles are congruent? Which sides correspond?

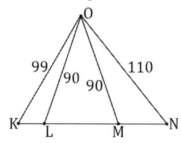

12. Consider the statement, "If two of the three interior angles are congruent for two triangles, the sides of the triangles come in the same proportions."

(A) Is this statement true?

(B) What is the converse of the given statement? Is the converse true?

(C) What is the inverse of the given statement? Is the inverse true?

(D) What is the contrapositive of the given statement? Is the contrapositive true?

Right Triangles

Right triangles are useful, even for studying some properties of acute triangles, obtuse triangles, and other shapes. Since many applications of geometry involve perpendicular lines, right triangles are common in subjects such as physics and engineering. Topics in this chapter include the Pythagorean theorem, the 30°-60°-90° triangle, the 45° right triangle, and how tests for congruence and similarity simplify for right triangles.

Chapter 5 Concepts

A **right** triangle has one interior angle with a measure of 90° (equivalent to $\frac{\pi}{2}$ rad). The **hypotenuse** is the longest side of a right triangle. The hypotenuse is opposite to the right angle. The **legs** are the shortest sides of a right triangle. The legs are opposite to the acute angles. For example, in the diagram below, c is the hypotenuse, a and b are the legs, and θ and φ are the acute angles. The two acute angles of a right triangle are complements. For example, $\theta + \varphi = 90°$ (equivalent to $\frac{\pi}{2}$ rad) in the triangle below.

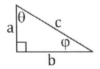

The **Pythagorean theorem**, which applies to all right triangles, states that the sum of the squares of the lengths of the two legs equals the square of the hypotenuse. For the triangle shown above, $a^2 + b^2 = c^2$ according to the Pythagorean theorem. If any two sides of a right triangle are known, the Pythagorean theorem (perhaps along with a little algebra) can be applied to solve for the remaining side.

Two special right triangles show up in a variety of math and science problems. One is the 30°-60°-90° triangle and the other is the 45°-45°-90° triangle.

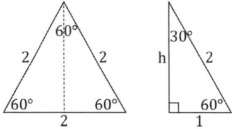

Recall from Chapter 2 that an equilateral triangle has three congruent sides and also has three 60° interior angles. If an equilateral triangle is cut in half as shown above, each small triangle is a 30°-60°-90° triangle. The edge length of the equilateral triangle above is 2 units. The hypotenuse of the 30°-60°-90° triangle above is 2 units because it is one side of the equilateral triangle. The base of the 30°-60°-90° triangle above is 1 unit because it is half as long as a side of the equilateral triangle; this side is opposite to the 30° angle. The height of the 30°-60°-90° triangle above can be found by applying the Pythagorean theorem. Let h represent the height of the 30°-60°-90° triangle.

$$1^2 + h^2 = 2^2$$
$$1 + h^2 = 4$$
$$h^2 = 4 - 1 = 3$$
$$h = \sqrt{3}$$

The height of the 30°-60°-90° triangle above is $\sqrt{3}$ units. The ratio of the sides of any 30°-60°-90° triangle is 1:$\sqrt{3}$:2, where the short side is opposite to the 30° angle and the long side is the hypotenuse. The hypotenuse of any 30°-60°-90° triangle is twice as long as the side opposite to the 30° angle. The side opposite to the 60° angle is $\sqrt{3}$ times as long as the side opposite to the 30° angle.

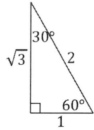

The unknown sides and angles of any 30°-60°-90° triangle can be found by using the ratio 1:$\sqrt{3}$:2.

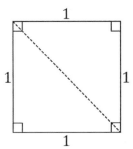
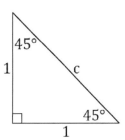

If a square is cut in half along its diagonal as shown above, this makes two 45°-45°-90° triangles. The edge length of the square above is 1 unit. Each leg of the 45°-45°-90° triangle above is 1 unit because it is one side of the square. Since each leg has the same length, the 45°-45°-90° triangle is an isosceles right triangle. The hypotenuse of the 45°-45°-90° triangle above can be found by applying the Pythagorean theorem. Let c represent the hypotenuse of the 45°-45°-90° triangle.

$$1^2 + 1^2 = c^2$$
$$1 + 1 = c^2$$
$$2 = c^2$$
$$\sqrt{2} = c$$

The hypotenuse of the 45°-45°-90° triangle above is $\sqrt{2}$ units. The ratio of the sides of any 45°-45°-90° triangle is 1:1:$\sqrt{2}$. The hypotenuse of any 45°-45°-90° triangle is $\sqrt{2}$ times as long as either of its legs.

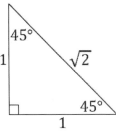

The unknown sides and angles of any 45°-45°-90° triangle can be found by using the ratio 1:1:$\sqrt{2}$.

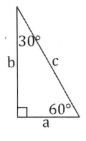
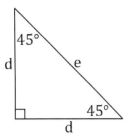

Consider the 30°-60°-90° triangle (left figure) and 45°-45°-90° triangle (right figure) on the previous page. The sides of the 30°-60°-90° triangle come in the ratio $1{:}\sqrt{3}{:}2$. These ratios lead to the following equations. Note that a is opposite to the 30° angle, b is opposite to the 60° angle, and c is the hypotenuse.

$$c = 2a$$
$$b = a\sqrt{3}$$
$$c\sqrt{3} = 2b$$

The sides of the 45°-45°-90° triangle come in the ratio $1{:}1{:}\sqrt{2}$. These ratios lead to the following equation. Note that d is the length of a leg and e is the hypotenuse.

$$e = d\sqrt{2}$$

Since some of the numbers will involve square roots, it may be useful to review a few algebra rules regarding square roots. For example, when a square root is squared, this effectively removes the radical sign.

$$\left(\sqrt{2}\right)^{2} = 2$$

When two square roots are multiplied together, the values inside get multiplied.

$$\sqrt{2}\sqrt{3} = \sqrt{2(3)} = \sqrt{6}$$

When two square roots are divided, the values inside get divided.

$$\frac{\sqrt{10}}{\sqrt{2}} = \sqrt{\frac{10}{2}} = \sqrt{5}$$

When an answer has a square root in the denominator, it is customary to multiply the numerator and denominator by the square root to **rationalize the denominator**.

$$\frac{1}{\sqrt{3}} = \frac{1}{\sqrt{3}}\frac{\sqrt{3}}{\sqrt{3}} = \frac{\sqrt{3}}{3}$$

When a number inside of a square root contains a factor that is a perfect square, it is customary to **factor the perfect square** out of the radical sign. For example, the number 18 can be factored as $18 = 9 \times 2$. The factor 9 is called a perfect square since $3^{2} = 9$.

$$\sqrt{18} = \sqrt{9(2)} = \sqrt{9}\sqrt{2} = 3\sqrt{2}$$

When working with right triangles, the tests for congruence and similarity (Chapters 3-4) can be simplified because one of the angles is 90°.

If two triangles are known to both be right triangles, following are ways to test if the two right triangles are **congruent**:

- If both legs are congruent, the right triangles are congruent. This is equivalent to SAS.
- If the hypotenuse is congruent and one leg is also congruent, the right triangles are congruent (**LH**). Since they are right triangles, the other leg must also be congruent by the Pythagorean theorem.
- If the hypotenuse is congruent and one acute angle is also congruent, the right triangles are congruent. This is equivalent to AAS.
- If one leg is congruent and one acute angle is also congruent, the right triangles are congruent. This is equivalent to ASA or AAS.

If two triangles are known to both be right triangles, following are ways to test if the two right triangles are **similar**:

- If one acute angle is congruent, the right triangles are similar. This is equivalent to AA.
- If the ratio of the hypotenuse to one leg is the same, the right triangles are similar (**LH similarity**).
- If the ratio of the legs is the same, the right triangles are similar. This is equivalent to SAS similarity.

Chapter 5 Examples

Example 1. Determine the hypotenuse for the triangle below.

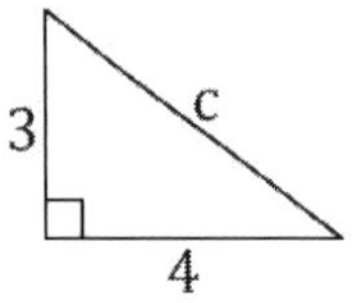

Apply the Pythagorean theorem.

$$3^2 + 4^2 = c^2$$
$$9 + 16 = c^2$$
$$25 = c^2$$

Square root both sides of the equation.

$$\sqrt{25} = c$$
$$5 = c$$

Example 2. Determine y for the triangle below.

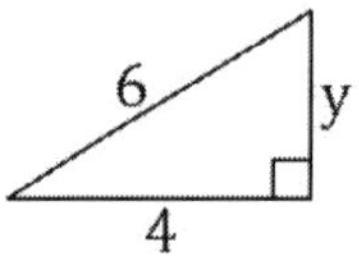

Apply the Pythagorean theorem.

$$4^2 + y^2 = 6^2$$
$$16 + y^2 = 36$$
$$y^2 = 36 - 16 = 20$$

Square root both sides of the equation.

$$y = \sqrt{20} = \sqrt{4(5)} = \sqrt{4}\sqrt{5} = 2\sqrt{5}$$

In the last step, a perfect square was factored out of the square root. Note that 4 is a perfect square (because $2^2 = 4$).

Example 3. Determine b and c for the triangle below.

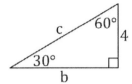

For a 30°-60°-90° triangle, the sides come in the ratio 1:$\sqrt{3}$:2. The side opposite to the 30° angle is the short side, the side opposite to the 60° angle is the middle side, and the hypotenuse is the long side. The ratio 4:b:c equals 1:$\sqrt{3}$:2. The hypotenuse (c) is twice the short side (4).

$$c = 2(4) = 8$$

The side opposite to the 60° angle (b) is $\sqrt{3}$ times longer than the side opposite to the 30° angle.

$$b = 4\sqrt{3}$$

Example 4. Determine a for the triangle below.

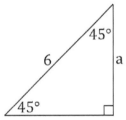

For a 45°-45°-90° triangle, the sides come in the ratio 1:1:$\sqrt{2}$. The hypotenuse (6) is $\sqrt{2}$ times longer than either leg (a).

$$a\sqrt{2} = 6$$

$$a = \frac{6}{\sqrt{2}}$$

Rationalize the denominator. Multiply the numerator and denominator each by $\sqrt{2}$. Recall from algebra that $\sqrt{2}\sqrt{2} = 2$.

$$a = \frac{6}{\sqrt{2}}\frac{\sqrt{2}}{\sqrt{2}} = \frac{6\sqrt{2}}{2} = 3\sqrt{2}$$

Example 5. Determine α and β in the diagram below.

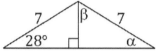

The two right triangles are congruent according to LH: The hypotenuses are congruent (both equal 7) and one leg is congruent (since the right triangles share the vertical leg). According to the CPCTC (Chapter 3), α = 28°. The two acute angles of a right triangle add up to 90°.

$$\alpha + \beta = 90°$$
$$28° + \beta = 90°$$
$$\beta = 90° - 28° = 62°$$

Example 6. Prove the Pythagorean theorem.

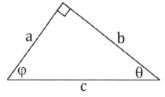 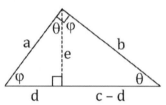

Any right triangle can be drawn with its hypotenuse as the base, like the left triangle above. A dashed vertical line can then be drawn from the top vertex to the base, which is perpendicular to the base. This divides the original right triangle into two smaller right triangles. The two acute angles, θ and φ, are complements. Since each smaller right triangle has θ or φ as one of its acute angles, the other two acute angles must be θ and φ, as labeled at the top. All three triangles (left, right, and original) are similar according to AA. The sides come in the same proportion. This means that a:b:c equals d:e:a and e:(c − d):b (in the order short:middle:hypotenuse). Note that a:c = d:a follows from a:b:c = d:e:a and that b:c = (c − d):b follows from a:b:c = e:(c − d):b.

$$\frac{a}{c} = \frac{d}{a} \quad \text{and} \quad \frac{b}{c} = \frac{c-d}{b}$$

Cross multiply. Multiply the left equation by ac and the right equation by bc.

$$a^2 = cd \quad \text{and} \quad b^2 = c^2 - cd$$

Add the equations together. The sum of the left-hand sides equals the sum of the right-hand sides. Note that cd cancels out: cd − cd = 0.

$$a^2 + b^2 = cd + c^2 - cd = c^2$$

Chapter 5 Problems

Note: The diagrams are not drawn to scale.

1. Determine the hypotenuse for the triangle below.

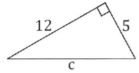

2. Determine a for the triangle below.

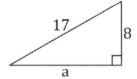

3. Determine d for the triangle below.

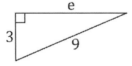

4. Determine e for the triangle below.

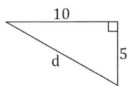

5. Determine h and θ for the triangle below.

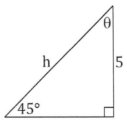

6. Determine b, c, and ∠2 for the triangle below.

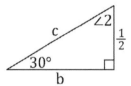

7. Determine a, e, and φ for the triangle below. The given angle is in radians.

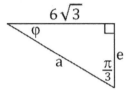

8. Determine d and ∠1 for the triangle below.

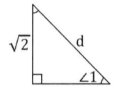

9. Determine t, w, and ∠4 for the triangle below.

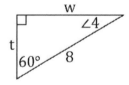

10. Determine m, n, and α for the triangle below. The given angle is in radians.

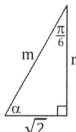

11. Determine k and θ for the triangle below.

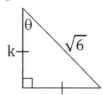

12. Show that the sides of a 30°-30°-120° triangle come in the ratio 1:1:√3. Hint: How can you make a right triangle?

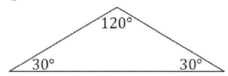

13. Prove that the two triangles below are congruent. Find p and q.

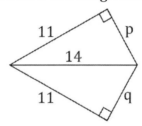

14. Prove that ΔABC~ΔBCD in the diagram below. Find AC and BD.

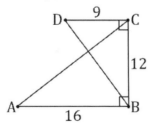

15. Prove that the two acute angles of any right triangle are complementary angles.

16. The surrounding triangle on the right below is congruent with the triangle at the left. On the right diagram, a square with edge length L was added so that its top right corner lies on the hypotenuse of the surrounding triangle. Prove that the three triangles (top, right, and surrounding) are similar. Prove that $\frac{1}{a} + \frac{1}{b} = \frac{1}{L}$.

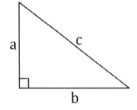 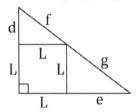

17. Consider the statement, "If a triangle has exactly two acute interior angles, it is a right triangle."

(A) Is this statement true?

(B) What is the converse of the given statement? Is the converse true?

(C) What is the inverse of the given statement? Is the inverse true?

(D) What is the contrapositive of the given statement? Is the contrapositive true?

Perimeter and Area of a Triangle

The area of a triangle is useful in geometry for multiple reasons. For one, the areas of quadrilaterals (Chapter 9) and other polygons (Chapter 10) can be found by dividing the shapes up into triangles. For another, the area of a triangle is sometimes useful in proofs or problems that do not seem to involve area. For example, some proofs of the Pythagorean theorem involve the formula for area.

Chapter 6 Concepts

The **perimeter** of a triangle equals the sum of the lengths of its sides. For example, the perimeter of the triangle below is $P = a + b + c$.

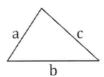

The **area** of a triangle equals one-half of the base times the height. Rotate the triangle until one side is at the bottom. The bottom side may be considered as the **base**. The **height** (which is also called **altitude**) is perpendicular to the base and passes through the opposite vertex. For example, the area of the surrounding triangle below is $A = \frac{1}{2}bh$.

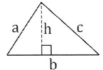

Any side of a triangle may be used as the base. The diagram on the next page illustrates the height corresponding to each possible base.

For an obtuse triangle, it is possible for the height to lie outside of the triangle, as the diagram below illustrates.

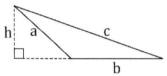

Heron's formula provides an alternative way to find the area of a triangle. Let a, b, and c represent the lengths of the three sides of a triangle. According to Heron's formula, the area of the triangle is

$$A = \sqrt{\frac{P}{2}\left(\frac{P}{2} - a\right)\left(\frac{P}{2} - b\right)\left(\frac{P}{2} - c\right)}$$

where $P = a + b + c$ is the perimeter of the triangle.

A 45°-45°-90° triangle is one-half of a square. Let L represent the length of each leg. The area of such a triangle is $\frac{L^2}{2}$.

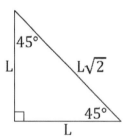

Recall from Chapter 5 that the sides of a 30°-60°-90° triangle come in the ratio $1:\sqrt{3}:2$. Let L represent the length of the side opposite to the 30° angle. The area of such a triangle is $\frac{1}{2}\left(L\sqrt{3}\right)L = \frac{L^2\sqrt{3}}{2}$.

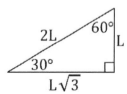

A 30°-60°-90° triangle is one-half of an equilateral triangle, as shown on the following page. The area of an equilateral triangle is twice the area of a 30°-60°-90° triangle. Let d represent the edge length of an equilateral triangle. Note that d = 2L, such that d/2 = L. Plug d/2 in for L in the formula for the area of a 30°-60°-90° triangle, and multiply by 2 since the area of an equilateral triangle is twice the area of a 30°-60°-90° triangle.

The area of the equilateral triangle below is $2\frac{L^2\sqrt{3}}{2} = L^2\sqrt{3} = \left(\frac{d}{2}\right)^2\sqrt{3} = \frac{d^2\sqrt{3}}{4}$.

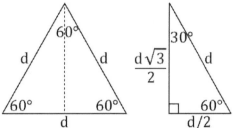

Note that it does not really matter what symbol is used. In the previous formulas, d is the edge length of the equilateral triangle, while 2L is the hypotenuse of the 30°-60°-90° half-triangle. With that notation, d = 2L, the area of the 30°-60°-90° half-triangle is $\frac{L^2\sqrt{3}}{2}$, and the area of the equilateral triangle is $\frac{d^2\sqrt{3}}{4}$.

If instead the symbol L is used for the side length of the equilateral triangle (rather than the hypotenuse of the 30°-60°-90° half-triangle), then the area of the equilateral triangle would be $\frac{L^2\sqrt{3}}{4}$. It is not important which letter is used in the formula. What is important is that the letter used in the formula be consistent with the letter used in diagrams or other formulas in the same problem. Consistency is key.

In the diagram below, the small square has area c^2, each right triangle has area $\frac{1}{2}ab$, and the large square has area $(a + b)^2 = a^2 + 2ab + b^2$. The area of the large square also equals the area of the small square plus four triangles: $c^2 + 4\left(\frac{1}{2}ab\right) = c^2 + 2ab$. Thus, $a^2 + 2ab + b^2 = c^2 + 2ab$, which proves the Pythagorean theorem: $a^2 + b^2 = c^2$.

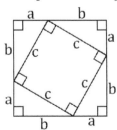

Chapter 6 Examples

Example 1. Determine the perimeter and area of the surrounding triangle below.

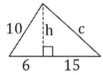

First use the Pythagorean theorem to find h.

$$6^2 + h^2 = 10^2$$

$$36 + h^2 = 100$$

$$h^2 = 100 - 36 = 64$$

$$h = \sqrt{64} = 8$$

Use the Pythagorean theorem again to find c.

$$h^2 + 15^2 = c^2$$

$$8^2 + 15^2 = c^2$$

$$64 + 225 = 289 = c^2$$

$$\sqrt{289} = 17 = c$$

The base of the surrounding triangle is b = 6 + 15 = 21. The perimeter of the surrounding triangle is P = 10 + b + c = 10 + 21 + 17 = 48. The height of the surrounding triangle is h = 8. The area of the surrounding triangle is:

$$A = \frac{1}{2}bh = \frac{1}{2}(21)(8) = \frac{1}{2}(168) = 84$$

Example 2. Use Heron's formula to find the area of the large triangle in Example 1. In Example 1, the perimeter was found to be P = 48. The three sides of the triangle are a = 10, b = 21, and c = 17.

$$A = \sqrt{\frac{P}{2}\left(\frac{P}{2} - a\right)\left(\frac{P}{2} - b\right)\left(\frac{P}{2} - c\right)} = \sqrt{\frac{48}{2}\left(\frac{48}{2} - 10\right)\left(\frac{48}{2} - 21\right)\left(\frac{48}{2} - 17\right)}$$

$$A = \sqrt{24(24 - 10)(24 - 21)(24 - 17)} = \sqrt{(24)(14)(3)(7)}$$

$$A = \sqrt{(3 \times 8)(2 \times 7)(3)(7)} = \sqrt{(3 \times 3)(7 \times 7)(8 \times 2)} = \sqrt{9}\sqrt{49}\sqrt{16} = 3(7)(4) = 84$$

The arithmetic was made simpler by regrouping the factors.

Example 3. Determine the perimeter and area of the triangle below.

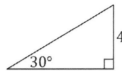

This is a 30°-60°-90° triangle. Recall from Chapter 5 that the sides of such a triangle come in the ratio $1:\sqrt{3}:2$. The height is $h = 4$, the base is $b = 4\sqrt{3}$, and the hypotenuse is $c = 8$. The perimeter is $P = h + b + c = 4 + 4\sqrt{3} + 8 = 12 + 4\sqrt{3} = 4(3 + \sqrt{3})$ and the area is $A = \frac{1}{2}bh = \frac{1}{2}(4\sqrt{3})(4) = 8\sqrt{3}$.

Example 4. Prove that the area of a right triangle equals one-half the product of its legs.

One way to make a right triangle is to cut a rectangle in half along its diagonal, as the diagram above illustrates. Since the opposite sides of a rectangle are congruent, the two right triangles are congruent according to SAS. The area of each right triangle is one-half of the area of the rectangle. Since the area of the rectangle is bh, the area of each right triangle is:

$$A = \frac{1}{2}bh$$

Example 5. Show that the area of ΔPQR below is one-half its base times its height.

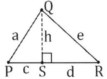

The height h divides ΔPQR into two right triangles. The area of ΔPQS is $\frac{1}{2}$ch. The area of ΔQRS is $\frac{1}{2}$dh. Add these together to find the area of ΔPQR.

$$A = \frac{1}{2}ch + \frac{1}{2}dh = \frac{1}{2}(c + d)h$$

Note that $(c + d)$ is the base of ΔPQR.

Chapter 6 Problems

Note: The diagrams are not drawn to scale.

1. Determine the perimeter and area of the triangle below.

2. Use Heron's formula to find the area of the triangle in Problem 1.

3. Determine the perimeter and area of the surrounding triangle below.

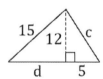

4. Use Heron's formula to find the area of the surrounding triangle in Problem 3.

5. Determine the perimeter and area of the obtuse triangle below.

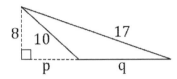

6. Use Heron's formula to find the area of the obtuse triangle in Problem 5.

7. Determine the area of the obtuse triangle below.

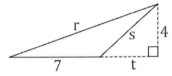

8. In the diagram below, points A, E, and I are collinear, points B, C, D, F, G, and H are collinear, $\overline{AI} \parallel \overline{BH}$, and BC $\cong$ DF $\cong$ GH. Rank the areas of the triangles in order from least to greatest.

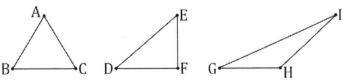

9. Determine the perimeter and area of the triangle below.

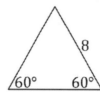

10. Determine the perimeter and area of the triangle below.

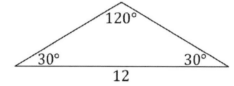

11. In the diagram below, BC ≅ CD ≅ DE and the area of ΔABD is 28. Determine the area of ΔABE.

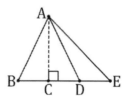

12. In the diagram below, DE = 3AB and the area of ΔABC is 12. Determine the area of ΔDEF.

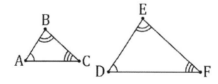

13. Show that the area of $\triangle XYZ$ below is one-half its base times its height.

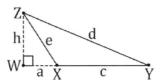

14. In the diagram below, $\triangle PQR$ is drawn twice. On the right, $\triangle PQR$ is divided into two right triangles. The parts of this problem relate to the diagram below.

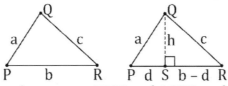

(A) Apply the Pythagorean theorem to $\triangle PQS$ and $\triangle QRS$, and use algebra to show that:

$$2bd = a^2 + b^2 - c^2$$
$$h^2 = a^2 - d^2$$

(B) Plug the answers to Part A into one-half the base times height for $\triangle PQR$ to show that:

$$A = \sqrt{\frac{4a^2b^2 - (a^2 + b^2 - c^2)^2}{16}}$$

(C) Show that the answer to Part B may be rewritten as:

$$A = \sqrt{\frac{(2ab + a^2 + b^2 - c^2)(2ab - a^2 - b^2 + c^2)}{16}}$$

(D) Show that the answer to Part C may be rewritten as:

$$A = \sqrt{\frac{[(a+b)^2 - c^2][c^2 - (a-b)^2]}{16}}$$

(E) Show that the answer to Part D may be rewritten as:

$$A = \sqrt{\frac{(a+b+c)(a+b-c)(a+c-b)(b+c-a)}{16}}$$

(F) Use the answer to Part E to derive Heron's formula.

15. Consider the statement, "If two triangles are congruent, they have the same area."

(A) Is this statement true?

(B) What is the converse of the given statement? Is the converse true?

(C) What is the inverse of the given statement? Is the inverse true?

(D) What is the contrapositive of the given statement? Is the contrapositive true?

Bisectors, Medians, and Altitudes

Some significant theorems in geometry relate to bisectors, medians, and altitudes. It is important to note that an angle bisector and a median are generally different, even though the definitions seem similar. The point where angle bisectors intersect, where medians intersect, or where altitudes intersect is often of importance. The point of intersection is generally different for angle bisectors, medians, and altitudes.

Chapter 7 Concepts

To **bisect** a shape means to cut the shape into two congruent parts. The **midpoint** of a line segment bisects the line segment. The midpoint is equidistant from the ends of the line segment. For example, point B is the midpoint of $\overline{AC}$ below. Point B bisects $\overline{AC}$, such that $\overline{AB} \cong \overline{BC}$. Point B is equidistant from points A and C.

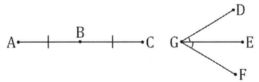

If a line segment is bisected, the line segment is cut into two shorter congruent line segments, as shown above on the left. An **angle bisector** is a line segment that cuts an angle into two congruent angles, as shown above on the right. Line segment $\overline{EG}$ in the right diagram is an angle bisector, such that $\angle DGE \cong \angle EGF$.

In a triangle, a **median** is a line segment that joins one vertex to the midpoint of the opposite side. For example, in the diagram that follows, point M is a midpoint and line segment $\overline{AM}$ is a median. Point M is equidistant from points B and C such that $\overline{BM} \cong \overline{MC}$. Although the median bisects $\overline{BC}$, it does NOT bisect ∠BAC in this example.

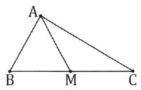

A **median** is generally different from an **angle bisector**. An angle bisector cuts an interior angle of a triangle in half, but generally does not bisect the side opposite to the angle. In the diagram below, $\overline{AM}$ is a median whereas $\overline{AD}$ is an angle bisector. In this example, the angle bisector and the median are two different line segments.

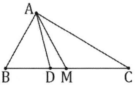

The only time that the median is also an angle bisector is if the triangle is **isosceles** and the median is drawn from the vertex where the two congruent sides intersect, as shown below. Since $\overline{AB} \cong \overline{AC}$, in this case $\overline{AM}$ is both a median and an angle bisector. In this case, ∠BAM ≅ ∠MAC and $\overline{BM} \cong \overline{MC}$.

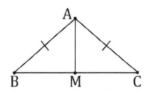

In the more general case, when the two sides of a triangle that form an interior angle are not congruent, the angle bisector is different from the median. According to the **triangle bisector theorem**, an angle bisector divides the opposite side of the triangle into segments in proportion to the lengths of the other two sides. For example, for the triangle shown below, $\overline{AD}$ is an angle bisector because it bisects ∠BAC such that ∠BAD ≅ ∠DAC. Since AB < AC, it follows that BD < DC. That is, point D is not a midpoint and AD is not a median. According to the triangle bisector theorem, if ∠BAD ≅ ∠DAC, it follows that $\frac{BD}{AB} = \frac{CD}{AC}$. Note that this is equivalent to $\frac{BD}{CD} = \frac{AB}{AC}$.

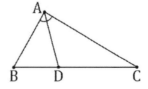

A **perpendicular bisector** passes perpendicularly through a line segment and cuts the line segment in half. A perpendicular bisector passes through the midpoint of the line segment that it is perpendicular to. For example, $\overline{BD}$ is a perpendicular bisector in the diagram below because $\overline{BD}$ is perpendicular to $\overline{AC}$ and bisects $\overline{AC}$. Note that $\overline{BD}$ passes through point M, which is the midpoint of $\overline{AC}$, such that $\overline{AM} \cong \overline{MC}$. (From the diagram below, it is not clear whether or not $\overline{AC}$ is the perpendicular bisector of $\overline{BD}$ because there are no tick marks to indicate whether or not $\overline{BM}$ is congruent with $\overline{MD}$.)

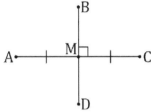

Any point that lies on a perpendicular bisector of a line segment is equidistant from the endpoints of the line segment. For example, every point that lies on $\overline{BD}$ above is equidistant from points A and C because $\overline{BD}$ is a perpendicular bisector of $\overline{AC}$.

For a line and a point that does not lie on that line, there exists exactly one line that is perpendicular to the given line and which passes through the given point. The shortest distance from the given point to the given line is along the line that is perpendicular to the given line. For example, in the diagram below, the shortest distance from point B to line $\overleftrightarrow{AC}$ is BD because $\overline{BD}$ is perpendicular to $\overleftrightarrow{AC}$.

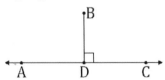

Any point that lies on an angle bisector is equidistant from the lines that form the angle. For example, since point K below lies on angle bisector $\overline{EH}$, point K is equidistant from $\overline{EG}$ and $\overline{EI}$. Since the shortest distance from a point to a line is perpendicular to the line, this means that $\overline{FK} \cong \overline{JK}$.

For a triangle, a **perpendicular bisector** passes perpendicularly through one side and cuts the side in half. For example, $\overline{DM}$ is a perpendicular bisector in the diagram below because $\overline{DM}$ is perpendicular to $\overline{BC}$ and bisects $\overline{BC}$. Note that $\overline{DM}$ passes through point M, which is the midpoint of $\overline{BC}$, such that $\overline{BM} \cong \overline{MC}$.

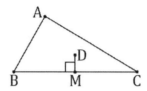

An **altitude** is a line segment that is perpendicular to one side of a triangle and connects to the vertex opposite to that side. (Altitude may also be referred to as height. The side that the altitude connects to may be considered the base. Rotate the triangle until the base is at the bottom. Then the term height will make sense.) The diagrams below draw three possible altitudes – one for each side. In each case, the dashed line segment is the altitude and the thick side is the side that it connects to.

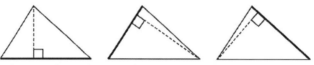

For an obtuse triangle, an altitude may lie outside of the triangle, as shown below.

Although the altitude is perpendicular to one side, the **altitude** is generally different from the **perpendicular bisector**. An altitude is only a perpendicular bisector if it cuts a side in half. The diagram below shows an altitude ($\overline{AD}$) and a perpendicular bisector ($\overline{EM}$) which are different. Altitude $\overline{AD}$ and perpendicular bisector $\overline{EM}$ are parallel. Any triangle has three altitudes (one for each vertex) and three perpendicular bisectors (one for each side). The altitude passes through a vertex (A), whereas the perpendicular bisector passes through a midpoint (M).

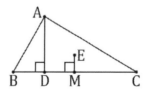

The diagram on the following page shows an angle bisector, a median, a perpendicular bisector, and an altitude for comparison. These four different special line segments relate to special properties of a triangle. The **angle bisector** ($\overline{AE}$) divides ∠BAC into two equal angles: ∠BAE ≅ ∠CAE. The **median** ($\overline{AM}$) divides side $\overline{BC}$ into two equal line segments and passes through vertex A: $\overline{BM}$ ≅ $\overline{CM}$. The **perpendicular bisector** ($\overline{DM}$) divides side $\overline{BC}$ into two equal line segments with a line that is perpendicular to side $\overline{BC}$: $\overline{BM}$ ≅ $\overline{CM}$. The **altitude** ($\overline{AF}$) is perpendicular to $\overline{BC}$ and passes through vertex A.

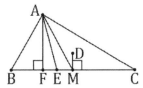

The three **angle bisectors** of any triangle are concurrent. (Recall from Chapter 1 that three lines are concurrent if they intersect at a single point.) The **incenter** is the name of the point where the three angle bisectors meet. The incenter is equidistant from the three sides of the triangle. (Volume 2 discusses the significance of the incenter.) In the diagram below, $\overline{AI}$, $\overline{BI}$, and $\overline{CI}$ are the three angle bisectors of ΔABC: $\overline{AI}$ bisects ∠BAC into ∠BAI ≅ ∠CAI, $\overline{BI}$ bisects ∠ABC into ∠ABI ≅ ∠CBI, and $\overline{CI}$ bisects ∠ACB into ∠ACI ≅ ∠BCI. Angle bisectors $\overline{AI}$, $\overline{BI}$, and $\overline{CI}$ are concurrent at the incenter (I). Line segments $\overline{DI}$, $\overline{EI}$, and $\overline{FI}$ are congruent and perpendicular to the sides of the triangle (but these are generally NOT perpendicular bisectors).

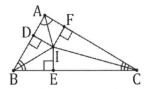

The three **perpendicular bisectors** of any triangle are concurrent. The **circumcenter** is the name of the point where the three perpendicular bisectors meet. The circumcenter is equidistant from the three vertices of the triangle. (Volume 2 discusses the significance of the circumcenter.) In the diagram below, $\overline{DO}$, $\overline{EO}$, and $\overline{FO}$ are the three perpendicular bisectors of ΔABC: $\overline{DO}$ bisects $\overline{AB}$ into $\overline{AD}$ ≅ $\overline{BD}$, $\overline{EO}$ bisects $\overline{BC}$ into $\overline{BE}$ ≅ $\overline{CE}$, and $\overline{FO}$ bisects $\overline{AC}$ into $\overline{AF}$ ≅ $\overline{CF}$. Perpendicular bisectors $\overline{DO}$, $\overline{EO}$, and $\overline{FO}$ are concurrent at the circumcenter (O). Line segments $\overline{AO}$, $\overline{BO}$, and $\overline{CO}$ are congruent.

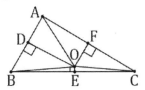

The three **medians** of any triangle are concurrent. The **centroid** is the name of the point where the three medians meet. The centroid cuts each median into line segments with a 2:1 ratio. In the diagram below, $\overline{AE}$, $\overline{BF}$, and $\overline{CD}$ are the three medians of ΔABC: $\overline{AE}$ bisects $\overline{BC}$ into $\overline{BE} \cong \overline{CE}$, $\overline{BF}$ bisects $\overline{AC}$ into $\overline{AF} \cong \overline{CF}$, and $\overline{CD}$ bisects $\overline{AB}$ into $\overline{AD} \cong \overline{BD}$. Medians $\overline{AE}$, $\overline{BF}$, and $\overline{CD}$ are concurrent at the centroid (G). The centroid (G) cuts medians $\overline{AE}$, $\overline{BF}$, and $\overline{CD}$ with a 2:1 ratio, such that AG = 2GE, BG = 2FG, and CG = 2DG. The centroid (G) is therefore one-third of the distance along each median from the corresponding midpoint (or two-thirds of the distance along each median from the corresponding vertex). For example, since AG = 2GE, point G is one-third of the length of AE from E to A. That is, GE = AE/3. Similarly, GF = BF/3 and DG = CD/3.

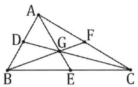

The three **altitudes** of any triangle are concurrent. The **orthocenter** is the name of the point where the three altitudes meet. (For an obtuse triangle, the orthocenter actually lies outside of the triangle. See Problem 22.) In the diagram below, $\overline{AE}$, $\overline{BF}$, and $\overline{CD}$ are the three altitudes of ΔABC: $\overline{AE}$ is perpendicular to $\overline{BC}$ and passes through A, $\overline{BF}$ is perpendicular to $\overline{AC}$ and passes through B, and $\overline{CD}$ is perpendicular to $\overline{AB}$ and passes through C. Altitudes $\overline{AE}$, $\overline{BF}$, and $\overline{CD}$ are concurrent at the orthocenter (H).

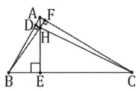

The **area** of a triangle equals one-half of its perimeter times its inradius: $A = \frac{1}{2}Pr$. The inradius is the shortest distance from the incenter to the sides of the triangle. (The names incenter and inradius will become clear in Volume 2.) For example, in the diagram below, r = DI = EI = FI is the inradius and $\overline{AI}$, $\overline{BI}$, and $\overline{CI}$ are the angle bisectors of ΔABC.

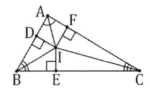

The three **medians** of a triangle divide the triangle into six smaller triangles which all have the same area. For example, medians $\overline{AE}$, $\overline{BF}$, and $\overline{CD}$ below divide $\triangle ABC$ into six triangles: $\triangle ADG$, $\triangle BDG$, $\triangle BEG$, $\triangle CEG$, $\triangle CFG$, and $\triangle AFG$. These six triangles have equal area. The area of one of these triangles is one-sixth of the area of $\triangle ABC$. Beware that although the areas of these six triangles are equal, these triangles are generally NOT congruent (in contrast to the four triangles discussed in the next paragraph).

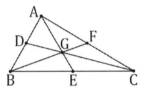

A **midsegment** (also called **midline**) of a triangle connects the midpoints of two sides. According to the **midsegment theorem** (or the **midline theorem**), each midsegment is parallel to one side of the triangle and is one-half as long as the side that it is parallel to. The three midsegments form four congruent triangles that are similar to the large triangle. For example, in $\triangle ABC$ below, D is the midpoint of $\overline{AB}$, E is the midpoint of $\overline{BC}$, and F is the midpoint of $\overline{AC}$. Midsegment $\overline{EF}$ is parallel to $\overline{AB}$, midsegment $\overline{DF}$ is parallel to $\overline{BC}$, and midsegment $\overline{DE}$ is parallel to $\overline{AC}$. Each midsegment is half the length of the side that it is parallel to: $EF = AB/2$, $DF = BC/2$, and $DE = AC/2$. The four small triangles ($\triangle ADF$, $\triangle BDE$, $\triangle DEF$, and $\triangle CEF$) are congruent. Each of the four small triangles is similar to $\triangle ABC$.

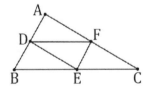

Chapter 7 Tips

Tips for working with angle bisectors:
- Each angle bisector cuts an interior angle into a pair of congruent angles.
- The three angle bisectors are concurrent at the incenter.
- The incenter is equidistant from the three sides of the triangle. The shortest distance from the incenter to any side is called the inradius.
- Line segments joining the incenter to each vertex and the three inradii divide the triangle into three pairs of congruent triangles.
- It may help to apply the triangle bisector theorem.
- The area of the triangle equals one-half of the perimeter times the inradius.

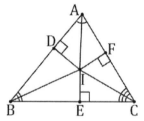

Tips for working with perpendicular bisectors:
- Each perpendicular bisector cuts one side of the triangle in half, passing through a midpoint.
- Each perpendicular bisector is perpendicular to one side of the triangle.
- The three perpendicular bisectors are concurrent at the circumcenter.
- The circumcenter is equidistant from the vertices of the triangle. The distance from the circumcenter to any vertex is called the circumradius.
- The perpendicular bisectors and three circumradii divide the triangle into three pairs of congruent triangles.
- It may help to apply the Pythagorean theorem to one of the smaller triangles.

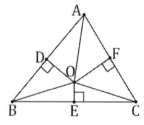

Tips for working with medians:

- Each median cuts one side of the triangle in half, passing through a midpoint.
- The three medians are concurrent at the centroid.
- The centroid lies one-third of the distance along a median from the midpoint. The distance from the centroid to a vertex is twice as long as the distance from the centroid to the corresponding midpoint.
- The medians form six triangles with equal area. Each triangle has one-sixth of the area of the surrounding triangle.
- A midsegment (or midline) joins two midpoints.
- Each midsegment is parallel to one side of the triangle. Each midsegment is one-half as long as the side that it is parallel to.
- The midsegments form four congruent triangles. Each triangle has one-fourth of the area of the surrounding triangle.

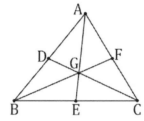 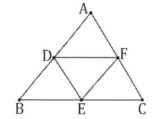

Tips for working with altitudes:

- Each altitude is perpendicular to one side of the triangle.
- Each altitude passes through one vertex.
- The three altitudes are concurrent at the orthocenter.
- The altitudes form three pairs of similar triangles.
- It may help to apply the Pythagorean theorem to one of the smaller triangles.
- It may help to make a ratio for a pair of triangles that are similar according to AA.
- The area of any triangle equals bh/2, with any side serving as the base.

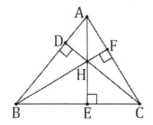

Chapter 7 Examples

Example 1. Determine AC in the diagram below.

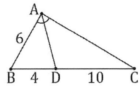

Since ∠BAD and ∠CAD are marked as congruent, the triangle bisector theorem may be applied. The ratio BD:AB equals the ratio CD:AC.

$$\frac{BD}{AB} = \frac{CD}{AC}$$

$$\frac{4}{6} = \frac{10}{AC}$$

Cross multiply. Multiply both sides of the equation by 6AC.

$$4AC = 6(10)$$

$$AC = \frac{60}{4} = 15$$

Check: BD/AB = 4/6 = 2/3 and CD/AC = 10/15 = 2/3. Alternatively, BD/CD = 4/10 = 2/5 and AB/AC = 6/15 = 2/5.

Example 2. In the diagram below, AG = 10, DG = 4, and DE = 7. Points D and E are midpoints of $\overline{AB}$ and $\overline{BC}$. Find EG, CG, and AC.

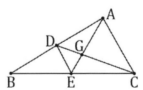

Since D and E are midpoints, $\overline{AE}$ and $\overline{CD}$ are medians, which makes point G the centroid, and $\overline{DE}$ is a midsegment. Since $\overline{AE}$ and $\overline{CD}$ are medians, AG = 2GE and CG = 2DG.

$$GE = \frac{AG}{2} = \frac{10}{2} = 5$$

$$CG = 2DG = 2(4) = 8$$

Since $\overline{DE}$ is a midsegment, AC = 2DE.

$$AC = 2DE = 2(7) = 14$$

Example 3. In the diagram below, $\angle ABC = 50°$ and $\angle ACB = 55°$. Line segments $\overline{AE}$, $\overline{BF}$, and $\overline{CD}$ are altitudes of $\triangle ABC$. Find $\angle BCD$, $\angle ACH$, and $\angle AHF$.

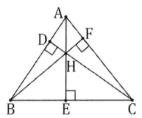

Since $\triangle BCD$ is a right triangle, $\angle CBD$ and $\angle BCD$ are complements. Note that $\angle CBD$ and $\angle ABC$ refer to the same angle.

$$\angle BCD = 90° - \angle CBD = 90° - \angle ABC = 90° - 50° = 40°$$

Note that $\angle BCD$ and $\angle ACH$ add up to $\angle ACB$.

$$\angle ACH = \angle ACB - \angle BCD = 55° - 40° = 15°$$

Since $\triangle ACE$ is a right triangle, $\angle ACE$ and $\angle CAE$ are complements. Note that $\angle ACE$ and $\angle ACB$ refer to the same angle.

$$\angle CAE = 90° - \angle ACE = 90° - \angle ACB = 90° - 55° = 35°$$

Since $\triangle AFH$ is a right triangle, $\angle AHF$ and $\angle HAF$ are complements. Note that $\angle CAE$ and $\angle HAF$ refer to the same angle.

$$\angle AHF = 90° - \angle HAF = 90° - \angle CAE = 90° - 35° = 55°$$

Example 4. In the diagram below, $BC = 6$, $AC = 4$, and $EO = 1$. Line segments $\overline{DO}$, $\overline{EO}$, and $\overline{FO}$ are perpendicular bisectors of $\triangle ABC$. Find FO. Note: B, O, and F are NOT collinear.

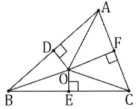

Since $\overline{EO}$ and $\overline{FO}$ are perpendicular bisectors, points E and F are midpoints. Therefore, $CE = BC/2 = 6/2 = 3$ and $CF = AC/2 = 4/2 = 2$. Use the Pythagorean theorem to find the hypotenuse of $\triangle CEO$: $EO^2 + CE^2 = CO^2$ becomes $1^2 + 3^2 = 1 + 9 = 10 = CO^2$, for which $\sqrt{10} = CO$. Use the Pythagorean theorem again with $\triangle CFO$: $CF^2 + FO^2 = CO^2$ becomes $2^2 + FO^2 = \left(\sqrt{10}\right)^2$, which simplifies to $4 + FO^2 = 10$. Subtract 4 from both sides of the equation: $FO^2 = 10 - 4 = 6$. Square root both sides: $FO = \sqrt{6}$.

Example 5. In the diagram below, D and E are midpoints. The area of ΔADE is 7. Find the area of ΔABC.

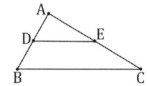

Since D and E are midpoints, $\overline{DE}$ is a midsegment and ΔADE is one of four congruent triangles that form when all three midsegments are drawn. The area of ΔADE is one-fourth of the area of ΔABC. The area of ΔABC is 4 × 7 = 28.

Example 6. In the diagram below, ΔABC is isosceles: $\overline{AB} \cong \overline{AC}$. Point M is the midpoint of $\overline{BC}$. Prove that $\overline{AM}$ is a median, angle bisector, altitude, and perpendicular bisector. Are the other two medians also angle bisectors, altitudes, and perpendicular bisectors?

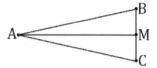

$\overline{AM}$ is a median because it connects a vertex (A) to the midpoint of the opposite side ($\overline{BC}$). Triangles ΔABM and ΔACM are congruent according to SSS: $\overline{AB} \cong \overline{AC}$ is given in the problem, $\overline{AM}$ is shared by the two triangles, and $\overline{BM} \cong \overline{CM}$ because point M is the midpoint of $\overline{BC}$. Since ΔABM ≅ ΔACM, ∠BAM ≅ ∠CAM and ∠AMB ≅ ∠AMC according to the CPCTC (Chapter 3). Since ∠BAM ≅ ∠CAM, $\overline{AM}$ is an angle bisector in addition to a median. Since ∠AMB and ∠AMC are congruent supplements, it follows that ∠AMB and ∠AMC are each right angles. Therefore, $\overline{AM}$ is an altitude and perpendicular bisector in addition to being an altitude and a median.

If $\overline{BC}$ is not congruent with $\overline{AB}$ and $\overline{AC}$, then the other two medians are NOT also angle bisectors, altitudes, or perpendicular bisectors.

This is a special property of isosceles triangles. For the vertex that touches both of the congruent sides, the median, angle bisector, altitude, and perpendicular bisector are the same.

Example 7. In the diagram below, point M is the midpoint of $\overline{BC}$. As shown in Example 6, $\overline{AM}$ is a median, angle bisector, altitude, and perpendicular bisector.

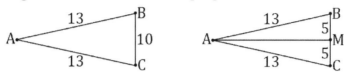

(A) How far is the centroid from point M?

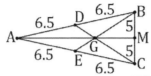

In the diagram above, D and E are the midpoints of $\overline{AB}$ and $\overline{AC}$. Line segments $\overline{AM}$, $\overline{CD}$, and $\overline{BE}$ are medians. Therefore, point G is the centroid, which means that GM = AG/2. Apply the Pythagorean theorem to determine AM.

$$BM^2 + AM^2 = AB^2$$
$$5^2 + AM^2 = 13^2$$
$$AM^2 = 169 - 25 = 144$$
$$AM = \sqrt{144} = 12$$

Since GM = AG/2, it follows that GM = AM/3 = 12/3 = 4. The centroid (G) is 4 units from point M. Note that AG = 2GM = 2(4) = 8 and that GM + AG = 4 + 8 = 12.

(B) How far is the incenter from point M? (This distance equals the inradius.)

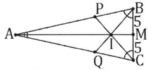

In the diagram above, $\overline{AM}$, $\overline{CP}$, and $\overline{BQ}$ are angle bisectors. Point I is the incenter. Apply the triangle bisector theorem to $\triangle ACM$. Since AM = 12 (from Part A), AI = 12 − IM.

$$\frac{AI}{AC} = \frac{IM}{CM}$$
$$\frac{12 - IM}{13} = \frac{IM}{5}$$
$$60 - 5IM = 13IM$$
$$60 = 18IM$$
$$IM = \frac{60}{18} = \frac{10}{3}$$

(C) How far is the circumcenter from point M? (This distance is NOT the circumradius.)

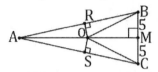

In the diagram above, $\overline{MO}$, $\overline{OR}$, and $\overline{OS}$ are perpendicular bisectors. (Neither $\overline{OR}$ nor $\overline{OS}$ is a median, but $\overline{MO}$ is a median according to Example 6.) Point O is the circumcenter. Since O is the circumcenter, $\overline{AO} \cong \overline{BO} \cong \overline{CO}$. Recall from Part A that AM = 12. It follows that AO = 12 − MO. Since $\overline{AO} \cong \overline{CO}$, it follows that CO = 12 − MO. Use the Pythagorean theorem for triangle $\triangle CMO$.

$$CM^2 + MO^2 = CO^2$$
$$5^2 + MO^2 = (12 - MO)^2$$
$$25 + MO^2 = 144 - 24MO + MO^2$$
$$24MO = 119$$
$$MO = \frac{119}{24}$$

(Note that the circumradius is the distance from O to a vertex, which is $12 - MO = \frac{169}{24}$.)

(D) How far is the orthocenter from point M?

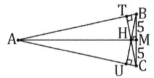

In the diagram above, $\overline{AM}$, $\overline{BU}$, and $\overline{CT}$ are altitudes. Point H is the orthocenter. Note that $\angle AHT \cong \angle CHM$ because these are vertical angles. Since $\triangle AHT$ and $\triangle CHM$ are both right triangles, $\angle HAT \cong \angle HCM$ since each angle is a complement to $\angle AHT \cong \angle CHM$. Note that $\angle BAM$ is the same as $\angle HAT$. Triangles $\triangle ABM$ and $\triangle CHM$ are similar according to AA (each has one right angle and an angle that is a complement to $\angle AHT \cong \angle CHM$). Since $\triangle ABM \sim \triangle CHM$, it follows that BM:AM:AB = HM:CM:CH.

$$\frac{BM}{AM} = \frac{HM}{CM}$$
$$\frac{5}{12} = \frac{HM}{5}$$
$$25 = 12HM$$
$$\frac{25}{12} = HM$$

Example 8. In the previous example, the inradius was found in Part B to be 10/3.

(A) Use the inradius to determine the area of ΔABC for the previous example.

The area of a triangle equals one-half of its perimeter times the inradius. The perimeter of ΔABC in the previous example is $P = 13 + 13 + 10 = 36$.

$$A = \frac{1}{2}Pr = \frac{1}{2}(36)\left(\frac{10}{3}\right) = (18)\left(\frac{10}{3}\right) = \frac{180}{3} = 60$$

(B) Show that the area from Part A agrees with the formula using base and height.

The area of a triangle equals one-half its base times its height. Taking $BC = 10$ to be the base, the corresponding height is $AM = 12$ (found in Part A of Example 7).

$$A = \frac{1}{2}bh = \frac{1}{2}(10)(12) = (5)(12) = 60$$

The formulas from Parts A and B of this example give the same value for the area (60).

Example 9. Prove that $\overline{BD}$ is the shortest possible segment that connects B to $\overleftrightarrow{AC}$ in the diagram below. Also prove that if $\overline{BD}$ is the perpendicular bisector for $\overline{AC}$, then B is equidistant from points A and C.

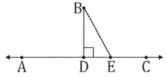

Let E be any point on $\overleftrightarrow{AC}$ other than D. Points B, D, and E form a right triangle. According to the Pythagorean theorem, $BE^2 = BD^2 + DE^2$. Since $DE > 0$, it follows that $BE > BD$. Therefore, BD is the shortest possible segment that connects B to $\overleftrightarrow{AC}$.

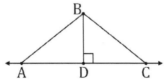

If $\overline{BD}$ is a perpendicular bisector for $\overline{AC}$, then $\overline{AD} \cong \overline{CD}$. According to SAS, $\Delta ABD \cong \Delta BCD$ (since side $\overline{BD}$ is shared and $\angle ADB = \angle CDB = 90°$). From the CPCTC (Chapter 3), it follows that $\overline{AB} \cong \overline{BC}$. (Alternatively, use the Pythagorean theorem to show that the two hypotenuses are congruent.)

Example 10. Show that the circumcenter of a triangle is equidistant from the vertices and show that the three perpendicular bisectors of a triangle are concurrent.

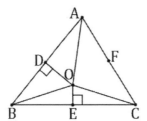

In the diagram above, points D, E, and F are midpoints, and line segments $\overline{DO}$ and $\overline{EO}$ are perpendicular bisectors. Let O be the point where $\overline{DO}$ and $\overline{EO}$ intersect. Since $\overline{DO}$ is the perpendicular bisector of $\overline{AB}$, point O must be equidistant from A and B (as shown in Example 9): $\overline{OA} \cong \overline{OB}$. Similarly, since $\overline{EO}$ is the perpendicular bisector of $\overline{BC}$, point O must be equidistant from B and C: $\overline{OB} \cong \overline{OC}$. Combine $\overline{OA} \cong \overline{OB}$ and $\overline{OB} \cong \overline{OC}$ to get $\overline{OA} \cong \overline{OB} \cong \overline{OC}$. This shows that point O is equidistant from vertices A, B, and C.

Since $\overline{OA} \cong \overline{OC}$, it follows that point O lies on the perpendicular bisector of $\overline{AC}$. Line segment $\overline{FO}$ (not drawn above) is therefore a perpendicular bisector of $\overline{AC}$. Since O is the point where $\overline{DO}$ and $\overline{EO}$ intersect, and point O also lies on $\overline{FO}$, it follows that point O is where all three of the perpendicular bisectors ($\overline{DO}$, $\overline{EO}$, and $\overline{FO}$) intersect. Point O is the circumcenter of ΔABC.

Chapter 7 Problems

Note: The diagrams are not drawn to scale.

1. In the diagram below, ∠BAC = 60°.

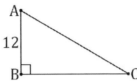

(A) Draw an angle bisector from vertex A to side $\overline{BC}$. Let D be the point on $\overline{BC}$ where this angle bisector intersects $\overline{BC}$. Determine BD.

(B) Draw a median from vertex A to side $\overline{BC}$. Let M be the point on $\overline{BC}$ where this median intersects $\overline{BC}$. Determine BM.

(C) Does the same line segment that bisects ∠BAC also bisect $\overline{BC}$? Explain.

2. In the diagram below, ∠ACB = 30°.

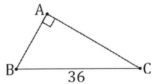

(A) Draw an altitude from vertex A to side $\overline{BC}$. Let D be the point on $\overline{BC}$ where this altitude intersects $\overline{BC}$. Determine BD.

(B) Draw a perpendicular bisector for side $\overline{BC}$. Let M be the point on $\overline{BC}$ where this perpendicular bisector intersects $\overline{BC}$. Determine BM.

(C) Find the distance between the altitude from Part A and the perpendicular bisector from Part B.

3. In the diagram below, ∠BAD ≅ ∠CAD. Determine AC.

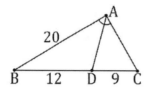

4. In the diagram below, ∠FEG ≅ ∠GEH. Determine FG.

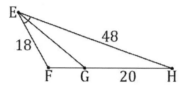

5. In the diagram below, ∠PRS ≅ ∠QRS and PQ = 21. Determine PS and QS.

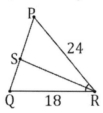

6. Determine VW and VX in the diagram below.

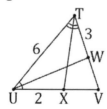

7. In the diagram below, DG = 10, BC = 24, and BG = 18. Points D and E are midpoints of $\overline{AB}$ and $\overline{AC}$. Find DE, EG, and CG.

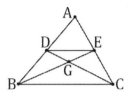

8. In the diagram below, PS = 36 and GR = 22. Points S and T are midpoints of $\overline{QR}$ and $\overline{PQ}$. Find GS, GT, and RT.

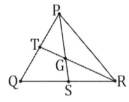

9. In the diagram below, AC = 16, DO = 5, and FO = 6. Line segments $\overline{DO}$, $\overline{EO}$, and $\overline{FO}$ are perpendicular bisectors of △ABC. Determine AO, AB, and BO.

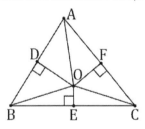

10. In the diagram below, LM = 2, NO = 2, and KP = $\sqrt{2}$. Line segments $\overline{NO}$, $\overline{OP}$, and $\overline{OQ}$ are perpendicular bisectors of △KLM. Determine MO, OP, and LO.

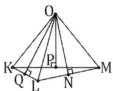

11. In the diagram below, AB = 10, CD = 12, and BE = 8. Line segments $\overline{BE}$ and $\overline{CD}$ are altitudes of ΔABC. Determine AC.

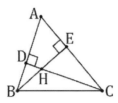

12. In the diagram below, HN = 6, LN = 8, and HP = 3. Line segments $\overline{KN}$ and $\overline{LP}$ are altitudes of ΔKLM. Find HL, HK, KN, KP, LP, MP, KM, MN, LM, and KL.

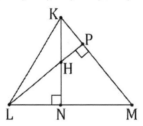

13. In the diagram below, ∠UTV = 48° and ∠TUV = 56°. Line segments $\overline{TX}$, $\overline{UY}$, and $\overline{VW}$ are angle bisectors of ΔTUV. Find α, δ, γ, θ, φ, η, ρ, σ, τ, β, μ, ν, ξ, ω, ψ, κ, λ, and χ. Are any triangles congruent or similar? If so, which ones? Are any inradii drawn? If so, which?

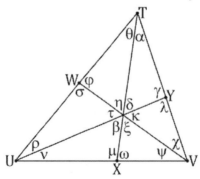

14. In the diagram below, ∠ABC = 64° and ∠ACB = 32°. Line segments $\overline{AI}$, $\overline{BI}$, and $\overline{CI}$ are angle bisectors of ΔABC. Find each of the 12 numbered angles. Are any triangles congruent or similar? If so, which ones? Which line segments are inradii?

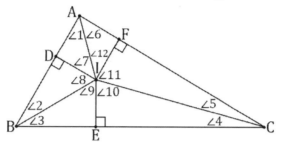

15. In the diagram below, $\overline{NO}$, $\overline{OP}$, and $\overline{OQ}$ are perpendicular bisectors of ΔKLM. Find α, δ, γ, θ, φ, β, μ, ν, κ, and λ. Are any triangles congruent or similar? If so, which ones? Which line segments are circumradii?

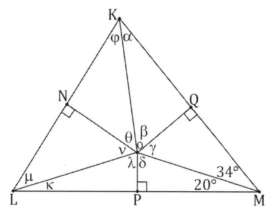

16. In the diagram below, ∠SRT = 72° and ∠RTS = 58°. Line segments $\overline{RV}$, $\overline{SW}$, and $\overline{TU}$ are altitudes of ΔRST. Find each of the 12 numbered angles. Are any triangles congruent or similar? If so, which ones?

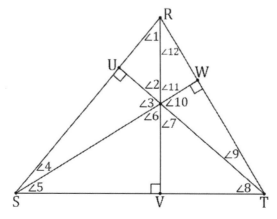

17. The equilateral triangle below has edge length L. Express each answer in terms of L.

(A) Find the length of an angle bisector, perpendicular bisector, median, and altitude.

(B) Find the shortest distance from the incenter to a side of the triangle.

(C) Find the distance from the circumcenter to a vertex.

(D) Find the height of the centroid relative to the base.

(E) Find the length of a midsegment.

(F) Find the area of the triangle using the base and height.

(G) Find the area of the triangle using the incenter and perimeter.

(H) Find the area of each of the six small triangles formed by the medians.

(I) Find the area of each of the four small triangles formed by the midsegments.

18. Each part of this problem refers to the triangle below with two congruent angles.

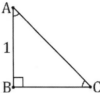

(A) State exactly where the orthocenter is and why. Find the lengths of the altitudes.

(B) State exactly where the circumcenter is and why. Find the circumradius.

(C) Draw the triangle with its three medians. Label the centroid as point G. Find BG.

(D) Draw the triangle with its three angle bisectors and the three shortest distances from the incenter to each side (similar to Problem 14, but NOT Problem 13). Find all of the angles in this diagram (similar to Problem 14).

(E) Label the incenter as point I. Find the inradius.

19. Each part of this problem refers to the triangle below.

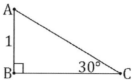

(A) State exactly where the orthocenter is and why. Find the lengths of the altitudes.

(B) State exactly where the circumcenter is and why. Find the circumradius.

(C) Draw the triangle with its three medians. Label the centroid as point G. Find BG.

(D) Draw the triangle with its three angle bisectors and the three shortest distances from the incenter to each side (similar to Problem 14, but NOT Problem 13). Find all of the angles in this diagram (similar to Problem 14).

(E) Find the inradius.

20. Each part of this problem refers to the triangle below.

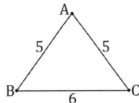

(A) Draw the triangle with its three altitudes. Label the orthocenter. Find the shortest distance from the orthocenter to $\overline{BC}$.

(B) Draw the triangle with its three perpendicular bisectors. Label the circumcenter. Find the circumradius.

(C) Draw the triangle with its three medians. Label the centroid. Find the shortest distance from the centroid to $\overline{BC}$.

(D) Draw the triangle with its three angle bisectors. Label the incenter. Find the inradius.

21. Each part of this problem refers to the triangle below.

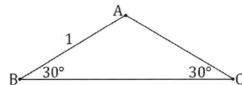

(A) Draw the triangle with its three altitudes. Label the orthocenter. Find the shortest distance from the orthocenter to $\overline{BC}$.

(B) Draw the triangle with its three perpendicular bisectors. Label the circumcenter. Find the circumradius.

(C) Draw the triangle with its three medians. Label the centroid. Find the shortest distance from the centroid to $\overline{BC}$.

(D) Draw the triangle with its three angle bisectors. Label the incenter. Find the inradius.

22. The left triangle below is acute. For each triangle below, indicate if the incenter, circumcenter, centroid, and orthocenter lie inside of the triangle, on the edge, or outside of the triangle. For any "center" that lies outside of a triangle, draw a diagram to show the approximate location of the center and the reason that it lies outside.

23. In the diagram below, points D, E, and F are midpoints of $\overline{AC}$, $\overline{BC}$, and $\overline{AB}$, and point G is the centroid of ΔABC. The area of ΔCDE is 18. Find the area of ΔABC, ΔADG, ΔACE, ΔDEF, ΔABG, and ΔDEG.

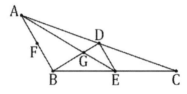

24. Find the area of the triangle below. Find the three altitudes of the triangle below. Determine the inradius. Hint: Recall Heron's formula.

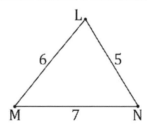

25. In the diagram below, $\overline{EH}$ bisects ∠GEI. Point K lies on angle bisector $\overline{EH}$. Prove that $\overline{FK} \cong \overline{JK}$.

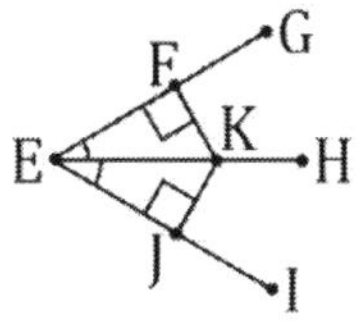

26. In the diagram below, $\overline{BC}$ bisects ∠ABE and $\overline{AC} \parallel \overline{BE}$. Use the diagram below to prove the triangle bisector theorem: DE/BE = AD/AB. Will the result of the proof remain valid if $\overline{AC}$ and $\overline{CD}$ are erased from the diagram? Explain.

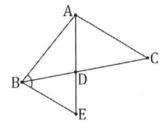

27. Prove that the orthocenter of any right triangle lies at the vertex where the two legs meet. Also prove that the circumcenter of any right triangle lies at the midpoint of the hypotenuse.

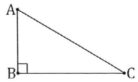

28. Show that the incenter of a triangle is equidistant from the sides and show that the three angle bisectors of a triangle are concurrent.

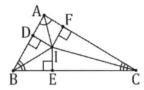

29. Show that the area of a triangle equals one-half the perimeter times the inradius.

30. In the diagram below, point D is the midpoint of $\overline{AB}$, point F is the midpoint of $\overline{AC}$, and $\overline{DF} \cong \overline{FP}$. Points A, F, and C are collinear, points D, F, and P are collinear, points A, D, and B are collinear, and points B, E, and C are collinear. Prove that $\overline{DF} \parallel \overline{BC}$ and that BC = 2DF.

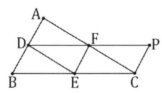

31. In the diagram below, D is the midpoint of $\overline{AB}$, E is the midpoint of $\overline{BC}$, and F is the midpoint of $\overline{AC}$. Prove that $\triangle ADF \cong \triangle CEF$.

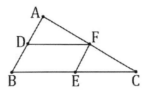

32. In the diagram below, D is the midpoint of $\overline{AB}$ and F is the midpoint of $\overline{AC}$, which means that G is the centroid. Prove that $\triangle DFG \sim \triangle BCG$, $BG = 2FG$, and $CG = 2DG$. Show that the three medians of a triangle are concurrent.

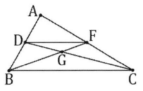

33. In the diagram below, $\overline{AE}$, $\overline{BF}$, and $\overline{CD}$ are medians of $\triangle ABC$. Show that the six triangles formed by the medians have equal area. Show that $\triangle ACG$, $\triangle BCG$, and $\triangle ABG$ have the same area.

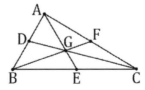

34. In the diagram below, $\overline{PQ} \parallel \overline{BC}$, $\overline{PR} \parallel \overline{AC}$, and $\overline{QR} \parallel \overline{AB}$. Show that the three altitudes of $\triangle ABC$ are concurrent.

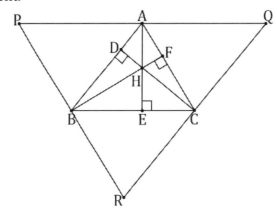

35. Consider the statement, "If a median is an angle bisector, at least two sides of the triangle are congruent."

(A) Is this statement true?

(B) What is the converse of the given statement? Is the converse true?

(C) What is the inverse of the given statement? Is the inverse true?

(D) What is the contrapositive of the given statement? Is the contrapositive true?

The Triangle Inequality

Given two sides of a triangle, the triangle inequality states which possible lengths the remaining side can have in order for the three sides to form a triangle. On a related note, the shortest side of a triangle is opposite to the smallest interior angle and the longest side of a triangle is opposite to the greatest interior angle. The Pythagorean theorem, which applies to right triangles, can be used to determine whether three given sides form an acute triangle, a right triangle, or an obtuse triangle.

Chapter 8 Concepts

According to the **triangle inequality**, the sum of the lengths of any two sides of a triangle is greater than the length of the remaining side. For example, for the triangle shown below, AB + AC > BC, AB + BC > AC, and AC + BC > AB. For an example with numbers, if two sides of a triangle are AB = 5 and BC = 7, the third side (AC) must be greater than 2 and less than 12. The reason that AC must be greater than 2 is so that AB + AC will be greater than BC. (This ensures that AB + AC will be greater than 7.) The reason that AC must be less than 12 is so that AB + BC will be greater than AC (since AB + BC = 12 exactly).

For any triangle, the shortest side is opposite to the smallest interior angle and the longest side is opposite to the greatest interior angle. For example, if $\overline{AC}$ is the longest side in the triangle above, then ∠ABC is the greatest interior angle (and the converse

---

is also true). If $\overline{AB}$ is the shortest side in the triangle on the previous page, then $\angle ACB$ is the smallest interior angle (and the converse is also true). If the lengths of the sides are ordered $AB < BC < AC$, then the angles are ordered $\angle ACB < \angle BAC < \angle ABC$ (and the converse is also true).

Let AB, BC, and AC be the three sides of a triangle, and let AC be the longest side: $AC > AB$ and $AC > BC$. If $AC^2 < AB^2 + BC^2$, the triangle is acute. If $AC^2 = AB^2 + BC^2$, the triangle is right. If $AC^2 > AB^2 + BC^2$, the triangle is obtuse. (The converse is also true for each case.) For example, consider the three triangles below. In each case, AC is the longest side: $AC > AB$ and $AC > BC$. The triangle on the left is an acute triangle: $AC^2 < AB^2 + BC^2$. The middle triangle is an obtuse triangle: $AC^2 > AB^2 + BC^2$. The triangle on the right is a right triangle: $AC^2 = AB^2 + BC^2$. For an example with numbers, suppose that $AB = 3$ and $BC = 4$ in each triangle below. In this case, $AB^2 + BC^2 = 3^2 + 4^2 = 9 + 16 = 25 = 5^2$. The triangle on the left is an acute triangle ($AC^2 < 5^2$), the middle triangle is an obtuse triangle ($AC^2 > 5^2$), and the triangle on the right is a right triangle ($AC^2 = 5^2$).

 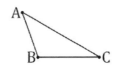

Chapter 8 Examples

Example 1. Two sides of a triangle have lengths of 6 and 10. What is the possible range of values for the length of the third side?

Let c represent the third side. The triangle inequality requires $6 + c > 10$, $10 + c > 6$, and $10 + 6 > c$. The first inequality simplifies to $c > 4$ (since $10 - 6 = 4$). The second inequality is automatically satisfied (since 10 is greater than 6). The third inequality simplifies to $16 > c$. Combine $c > 4$ with $16 > c$ together: $16 > c > 4$. Note that this is equivalent to $4 < c < 16$. The third side must be greater than 4 and less than 16.

Example 2. Order sides a, b, and c in the diagram below from longest to shortest.

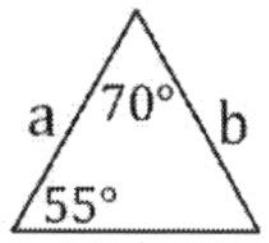

First, solve for the third interior angle, knowing that the three interior angles of any triangle add up to 180°.

$$70° + 55° + \angle 3 = 180°$$
$$\angle 3 = 180° - 70° - 55° = 55°$$

Since 70° is the largest interior angle, c is the longest side (since it is opposite to 70°). Since the other two interior angles both equal 55°, a and b are the same length. The longest side is c and sides a and b tie for the shortest side. It turns out that this triangle is isosceles.

Example 3. The lengths of the sides of a triangle are 4, 7, and 9. Is this triangle acute, right, obtuse, or impossible?

Since the longest side is 9, let $AC = 9$, $AB = 4$, and $BC = 7$. Add the squares of the two shortest sides and compare this sum to the square of the longest side. Since $AC^2 = 9^2$ $= 81$ is greater than $AB^2 + BC^2 = 4^2 + 7^2 = 16 + 49 = 65$, the triangle is obtuse. (To tell whether or not the triangle is possible, check if it satisfies the triangle inequality: $9 + 4 > 7$, $9 + 7 > 4$, and $7 + 4 > 9$.)

Example 4. In the triangle below, AB > AC. Prove that ∠ACB > ∠ABC.

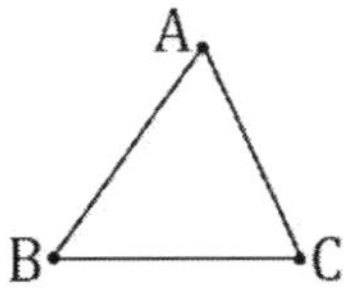

Draw altitude $\overline{AD}$. Add point E to the right such that $\overline{AE}$ is the mirror image of $\overline{AB}$. In the diagram below, $\overline{AB} \cong \overline{AE}$, which makes ΔABE an isosceles triangle. Recall that the interior angles opposite to the congruent sides of an isosceles triangle are congruent: ∠ABE ≅ ∠AEB. Point E must be to the right of point C because AE = AB > AC. (Use the Pythagorean theorem to prove this. Since $AD^2 + DE^2 = AE^2$ and $AD^2 + CD^2 = AC^2$, it follows from AE > AC that DE > CD.) Since ∠ACE and ∠AEC are two interior angles of ΔACE and since all three interior angles of any triangle add up to 180°, it follows that 180° > ∠ACE + ∠AEC. Since ∠ACB and ∠ACE are supplements, ∠ACE = 180° − ∠ACB. Replace ∠ACE with 180° − ∠ACB in the previous inequality.

$$180° > 180° - \angle ACB + \angle AEC$$
$$0 > -\angle ACB + \angle AEC$$
$$\angle ACB > \angle AEC$$

Since ∠AEC is the same as ∠AEB and since ∠ABC ≅ ∠AEB, this inequality is equivalent to ∠ACB > ∠ABC.

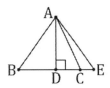

Chapter 8 Problems

Note: The diagrams are not drawn to scale.

1. Each pair of numbers below includes the lengths of two sides of a triangle. For each pair, indicate the possible range of values for the length of the third side.

(A) 3 and 5   (B) 6 and 6

(C) 2 and 9   (D) 12 and 21

2. Order sides a, b, and c in the diagram below from longest to shortest.

3. Each set of numbers below includes the lengths of three sides of a triangle. For each set, indicate whether the triangle is acute, right, or obtuse, or if it would be impossible for these lengths to be the sides of a triangle.

(A) 6, 7, 9   (B) 6, 12, 14

(C) 8, 15, 17   (C) 9, 12, 24

4. In the diagram below, $\overline{AD}$ is an altitude of $\triangle ABC$. Prove that $AB > BD$ and $AC > DC$. Use these inequalities to prove that $AB + AC > BC$.

5. Let AB, BC, and AC be the three sides of a triangle, and let AC be the longest side. Show that $AC^2 < AB^2 + BC^2$ if the triangle is acute, $AC^2 = AB^2 + BC^2$ if the triangle is right, and $AC^2 > AB^2 + BC^2$ if the triangle is obtuse.

 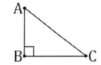 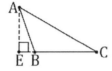

6. The three edge lengths of a triangle are AB, BC, and AC. Without knowing which of the edges is longest or shortest, prove that $AB > |BC - AC|$, where $|BC - AC|$ means to find the absolute value of the difference.

7. Consider the statement, "If $AC^2 > AB^2 + BC^2$, $\angle ABC$ is an obtuse angle." Note: A, B, and C represent the three vertices of a triangle.

(A) Is this statement true?

(B) What is the converse of the given statement? Is the converse true?

(C) What is the inverse of the given statement? Is the inverse true?

(D) What is the contrapositive of the given statement? Is the contrapositive true?

Quadrilaterals

This chapter explores properties of quadrilaterals, including parallelograms, trapezoids, and kites. Special parallelograms include squares, rectangles, and rhombuses.

Chapter 9 Concepts

A **quadrilateral** is a closed plane figure bounded by four straight sides. A **parallelogram** is a quadrilateral that has two pairs of parallel edges. Examples of parallelograms are shown below. A **rhombus** (second figure from the left below) is a parallelogram that is **equilateral** (it has four congruent sides). A **rectangle** (third figure from the left) is a parallelogram that is **equiangular** (it has four congruent interior angles). A **square** (right figure) is a parallelogram that is **regular** (it is both equilateral and equiangular).

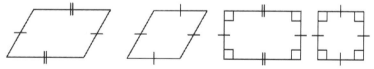

A **trapezoid** is a quadrilateral that has one pair of parallel edges (left figure below). A **kite** has two pairs of congruent edges, but does not have any parallel edges (right figure below). Trapezoids and kites are quadrilaterals that are NOT parallelograms.

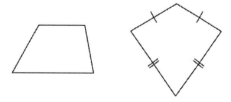

The four **interior angles** form at the four vertices and lie inside of the quadrilateral. For the quadrilateral below, the interior angles are ∠BAD, ∠ABC, ∠BCD, and ∠ADC. An **exterior angle** forms between one side of the quadrilateral and a line that extends from an adjacent side. Below, α is an interior angle while β is an exterior angle.

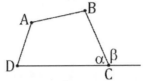

The sum of the **interior angles** of any quadrilateral equals 360° (which is equivalent to 2π rad). For example, ∠BAD + ∠ABC + ∠BCD + ∠ADC = 360° for the quadrilateral above. The sum of the **exterior angles** of any quadrilateral also equals 360°. For example, ∠1 + ∠2 + ∠3 + ∠4 = 360° for the quadrilateral below.

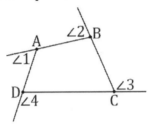

The two pairs of opposite sides of a **parallelogram** are parallel and congruent. Any two consecutive interior angles of a parallelogram are supplementary. One way to show that a quadrilateral is a parallelogram is to show that both pairs of opposite sides are congruent. One alternative is to show that one pair of opposite sides are both parallel and congruent. Another alternative is to show that both pairs of opposite angles are congruent. Yet another way is to show that the diagonals bisect each other's lengths.

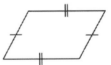

A **rhombus** is an **equilateral** parallelogram. All four sides of a rhombus are congruent. The **diagonals** of a rhombus are perpendicular and bisect the interior angles. One way to show that a quadrilateral is a rhombus is to show that all four edges are congruent. One alternative is to show that it is a parallelogram with perpendicular diagonals.

A **rectangle** is an **equiangular** parallelogram. All four interior angles of a rectangle are right angles. The **diagonals** of a rectangle are congruent. One way to show that a quadrilateral is a rectangle is to show that three of the interior angles are right angles. One alternative is to show that it is a parallelogram with congruent diagonals.

A **square** is a **regular** parallelogram. A square is both equilateral and equiangular. All four sides of a square are congruent and all four interior angles of a square are right angles. The **diagonals** of a square are both congruent and perpendicular. In order for a quadrilateral to be a square, it must satisfy the criteria for a rhombus and rectangle both.

The **diagonals** of a **parallelogram** bisect each other's lengths. The sum of the squares of the diagonals equals the sum of the squares of the sides. For example, $AC^2 + BD^2 = AB^2 + BC^2 + CD^2 + AD^2 = 2CD^2 + 2BC^2$ for the parallelogram below.

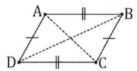

One pair of opposite sides of a **trapezoid** are parallel. The parallel sides of a trapezoid are called the **bases**. The other sides of a trapezoid are called the **legs**. For example, in the trapezoid below, AB and CD are the bases while AD and BC are the legs. The **median** is the line segment that connects the midpoints of the legs. The median is parallel to the bases. The length of the median equals the average length of the bases. For example, E and F are the midpoints of $\overline{AD}$ and $\overline{BC}$ in the diagram below, and $\overline{EF}$ is the median. The length of the median is $EF = \frac{AB+CD}{2}$.

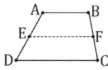

In an **isosceles trapezoid**, the two legs are congruent, there are two pairs of congruent base angles, and the two diagonals are congruent. For example, $\overline{AD} \cong \overline{BC}$, $\angle BAD \cong \angle ABC$, $\angle ADC \cong \angle BCD$, and $\overline{AC} \cong \overline{BD}$ in the isosceles trapezoid on the following page.

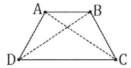

Two pairs of adjacent sides are congruent for a **kite**. The **diagonals** of a kite are perpendicular. A kite has one pair of congruent opposite interior angles. For example, $\overline{AD} \cong \overline{AB}$, $\overline{CD} \cong \overline{BC}$, $\angle ADC \cong \angle ABC$, and $\overline{AC} \perp \overline{BD}$ for the kite below.

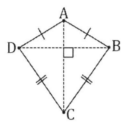

The **perimeter** of a **quadrilateral** equals the sum of its edge lengths: $P = AB + BC + CD + AD$. For a **parallelogram**, this simplifies to doubling each different side and adding: $P = 2L_1 + 2L_2$. For a rectangle, this becomes twice the length plus twice the width: $P = 2L + 2W$. The **area** of a **parallelogram** equals its base times its height: $A = bh$. The area of a **rhombus** can be found as $A = bh$ or as one-half the product of its diagonals: $A = \frac{1}{2}d_1d_2$. The area of a **rectangle** is $A = LW$ and the area of a **square** is $A = L^2$. The area of a **trapezoid** is one-half the sum of its bases times the height: $A = \frac{b_1 + b_2}{2}h$, which is equivalent to the median times the height: $A = mh$. The area of a **kite** is one-half of the product of its diagonals: $A = \frac{1}{2}d_1d_2$.

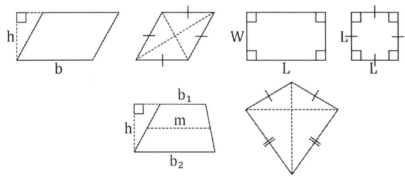

The four vertices are used to refer to the name of a quadrilateral. For example, MNOP refers to a quadrilateral with vertices M, N, O, and P.

Chapter 9 Examples

Example 1. Determine $\angle BAD$ in the diagram below.

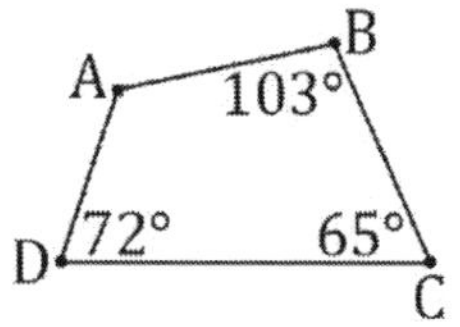

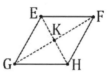

The interior angles of a quadrilateral add up to $360°$.

$$\angle BAD + 103° + 65° + 72° = 360°$$
$$\angle BAD + 240° = 360°$$
$$\angle BAD = 360° - 240° = 120°$$

Example 2. In rhombus EFGH below, $EH = 6$ and $FG = 8$. Determine the edge lengths, perimeter, and area of the rhombus.

The diagonals of any parallelogram (including the rhombus) bisect each other: $EK = HK = EH/2 = 6/2 = 3$ and $GK = FK = FG/2 = 8/2 = 4$. The diagonals of a rhombus are perpendicular: $\overline{EH} \perp \overline{FG}$. This means that the diagonals divide the rhombus into four right triangles. Each right triangle has legs equal to 3 and 4. The hypotenuse of each triangle is one side of the rhombus. Apply the Pythagorean theorem to one of these right triangles: $EK^2 + FK^2 = EF^2$. Plug in $EK = 3$ and $FK = 4$.

$$3^2 + 4^2 = EF^2$$
$$9 + 16 = EF^2$$
$$25 = EF^2$$
$$\sqrt{25} = 5 = EF$$

Since EFGH is a rhombus, all four sides are congruent with edge length 5. The perimeter of the rhombus is $P = EF + FH + GH + EG = 5 + 5 + 5 + 5 = 20$. The area of the rhombus equals one-half the product of its diagonals.

$$A = \frac{1}{2} d_1 d_2 = \frac{1}{2}(EH)(FG) = \frac{1}{2}(6)(8) = 24$$

Example 3. In the diagram below, AB = 16, DE = 6, CE = 18, and AE = 8. Determine the median and area for trapezoids ABCD and ABCE.

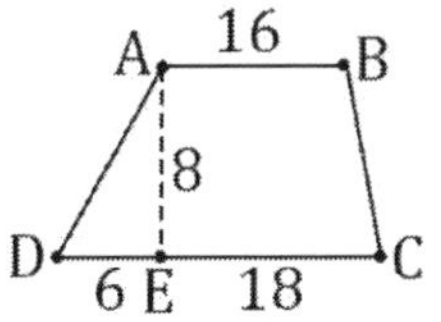

For trapezoid ABCD, the bases are CD = 6 + 18 = 24 and AB = 16. The median is the average of the bases.

$$m = \frac{AB + CD}{2} = \frac{16 + 24}{2} = \frac{40}{2} = 20$$

The area of trapezoid ABCD is its median times its height: A = mh = (20)(8) = 160.

For trapezoid ABCE, the bases are CE = 18 and AB = 16. The median is the average of the bases.

$$m = \frac{AB + CE}{2} = \frac{16 + 18}{2} = \frac{34}{2} = 17$$

The area of trapezoid ABCE is its median times its height: A = mh = (17)(8) = 136.

Example 4. A rectangle is twice as long as it is wide and has an area of 32. What is the perimeter of the rectangle?

The length is twice the width: L = 2W. The area of a rectangle equals its length times its width: A = LW. Replace L with 2W in the formula for area: $A = (2W)W = 2W^2$. The area equals 32.

$$32 = 2W^2$$
$$16 = W^2$$
$$\sqrt{16} = 4 = W$$
$$L = 2W = 2(4) = 8$$

The width of the rectangle is W = 4 and the length of the rectangle is L = 8. Plug these values into the formula for the perimeter of a rectangle.

$$P = 2L + 2W = 2(8) + 2(4) = 16 + 8 = 24$$

Example 5. Prove that the diagonals of a rhombus are perpendicular.

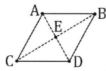

A rhombus is a parallelogram, which means that $\overline{AB} \parallel \overline{CD}$ and $\overline{AC} \parallel \overline{BD}$. A rhombus is an equilateral quadrilateral, which means that $\overline{AB} \cong \overline{CD} \cong \overline{AC} \cong \overline{BD}$. The diagonals of a parallelogram bisect each other's lengths, which means that $AE = DE = AD/2$ and $CE = BE = BC/2$. $\triangle ACE \cong \triangle ABE$ according to SSS because $\overline{AB} \cong \overline{AC}$, $\overline{CE} \cong \overline{BE}$, and $\overline{AE}$ is shared by $\triangle ACE$ and $\triangle ABE$. From the CPCTC, it follows that $\angle AEC \cong \angle AEB$. These angles are supplementary angles: $\angle AEC + \angle AEB = 180°$. Replace $\angle AEB$ with $\angle AEC$ to get $\angle AEC + \angle AEC = 2\angle AEC = 180°$. Divide by 2 to get $\angle AEC = 90°$, which shows that $\overline{AD} \perp \overline{BC}$.

Example 6. Prove that the sum of the exterior angles of a quadrilateral is 360°.

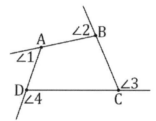

The sum of the interior angles of a quadrilateral is 360°.

$$\angle BAD + \angle ABC + \angle BCD + \angle ADC = 360°$$

Each exterior angle is a supplement to one of the interior angles.

$$\angle 1 + \angle BAD = 180° \quad , \quad \angle 2 + \angle ABC = 180°$$
$$\angle 3 + \angle BCD = 180° \quad , \quad \angle 4 + \angle ADC = 180°$$

Subtract the exterior angle from both sides of each equation.

$$\angle BAD = 180° - \angle 1 \quad , \quad \angle ABC = 180° - \angle 2$$
$$\angle BCD = 180° - \angle 3 \quad , \quad \angle ADC = 180° - \angle 4$$

Substitute these expressions into the sum of the interior angles formula.

$$180° - \angle 1 + 180° - \angle 2 + 180° - \angle 3 + 180° - \angle 4 = 360°$$

Add the exterior angles to both sides and subtract 360° from both sides.

$$180° + 180° + 180° + 180° - 360° = \angle 1 + \angle 2 + \angle 3 + \angle 4$$
$$360° = \angle 1 + \angle 2 + \angle 3 + \angle 4$$

Chapter 9 Problems

Note: The diagrams are not drawn to scale.

1. Determine ∠1 in the diagram below.

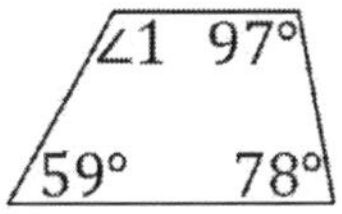

2. In parallelogram ABCD below, ∠ADF = 60°, DF = 5, and CF = 15. Determine ∠DAE, ∠ABC, ∠BCD, AD, AB, BC, and the height, perimeter, and area of parallelogram ABCD.

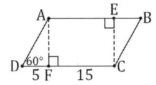

3. Find the median, perimeter, and area of trapezoid GHIJ below.

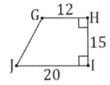

4. In parallelogram KLMN below, LM = 13, MP = 5, and NP = 18. Determine the height, perimeter, and area of parallelogram KLMN.

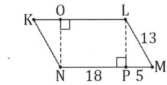

5. Find ∠BAD, the height, median, diagonals, perimeter, and area of isosceles trapezoid ABCD below.

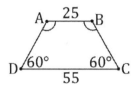

6. In rhombus EFGH below, EH = 10 and EG = 12. Find the perimeter and area of EFGH.

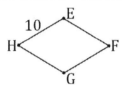

7. Find the perimeter and area of quadrilateral IJKL below.

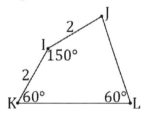

8. Find the perimeter and area of kite MNOP below.

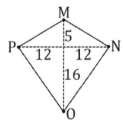

9. The perimeter of a square is 72. What is the area of the square?

10. The perimeter of a rectangle is 56. The area of the rectangle is 192. What are the length and width of the rectangle? What are the diagonals of the rectangle?

11. Prove that the interior angles of an equiangular quadrilateral are right angles.

12. Use the diagram below to prove that the four interior angles of a quadrilateral add up to 360°.

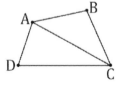

13. Prove that if the two diagonals of a quadrilateral bisect each other's lengths, then the quadrilateral is a parallelogram.

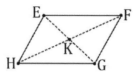

14. Prove that if both pairs of opposite sides of a quadrilateral are congruent, then the quadrilateral is a parallelogram.

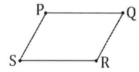

15. Prove that any two adjacent interior angles of a parallelogram are supplementary.

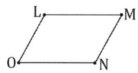

16. Prove that if both pairs of opposite interior angles of a quadrilateral are congruent, then the quadrilateral is a parallelogram.

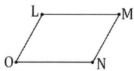

17. Prove that if the diagonals of a parallelogram are congruent, then the parallelogram is a rectangle.

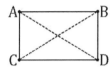

18. Prove that $PR^2 + QS^2 = 2RS^2 + 2QR^2$ for the parallelogram below.

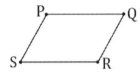

19. Prove that the diagonals of a rhombus are angle bisectors.

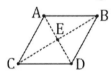

20. Prove that diagonal $\overline{TW}$ of kite UTVW below bisects ∠UTV and ∠UWV.

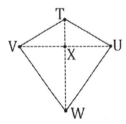

21. Prove that the diagonals of a kite are perpendicular.

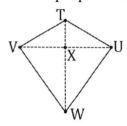

22. Show that the area of a parallelogram equals its base times its height: $A = bh$.

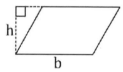

23. Show that the area of a rhombus equals one-half the product of its diagonals: $A = \frac{1}{2}d_1d_2$.

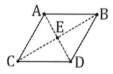

24. Show that the area of a trapezoid is one-half the sum of its bases times the height: $A = \frac{a+b}{2}h$.

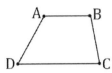

25. Show that the area of a kite is one-half of the product of its diagonals: $A = \frac{1}{2}d_1d_2$.

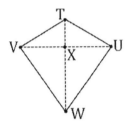

26. Consider the statement, "If a shape is a parallelogram, it is a quadrilateral."

(A) Is this statement true?

(B) What is the converse of the given statement? Is the converse true?

(C) What is the inverse of the given statement? Is the inverse true?

(D) What is the contrapositive of the given statement? Is the contrapositive true?

Polygons

This chapter covers a variety of properties of polygons. Topics include convex and concave polygons, reflex angles, regular polygons, equiangular and equilateral polygons, and sums of interior and exterior angles.

Chapter 10 Concepts

A **polygon** is a closed plane figure bounded by straight sides (also called edges). The number of sides equals the number of vertices. A polygon is named after the number of sides. A **triangle** has three sides, a **quadrilateral** has four sides, a **pentagon** has five sides, a **hexagon** has six sides, a **heptagon** has seven sides, an **octagon** has eight sides, a **nonagon** has nine sides, a **decagon** has ten sides, a **hendecagon** (or undecagon) has eleven sides, and a **dodecagon** has twelve sides.

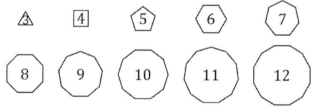

An **interior** angle is formed by two sides that meet at a vertex of a polygon, and lies inside of the polygon. An **exterior** angle is formed by one side of a polygon and a line that extends from an adjacent side. For example, θ is an interior angle while φ is an exterior angle in the diagram below.

In a **convex** polygon, every interior angle is less than 180°. A **concave** polygon has at least one **reflex angle** as an interior angle. (Recall from Chapter 1 that a reflex angle is an angle that is greater than 180°.) In a concave polygon, at least one vertex appears to be pushed inward, like polygon GHIJ below. For a concave polygon, it is possible to draw a line that intersects the boundary of the polygon at more than two points, and it is possible to draw a diagonal that lies outside of the polygon. It is not possible for a triangle to be concave, but a concave polygon is possible with more than three sides. Polygon ABCD below is convex, whereas polygon GHIJ below is concave. Note that line segment $\overline{KL}$ intersects the boundary of concave quadrilateral GHIJ in four different places, whereas line segment $\overline{EF}$ only intersects the boundary of convex quadrilateral ABCD at two different places. Also, diagonal $\overline{GI}$ lies outside of GHIJ, while every diagonal of ABCD, including $\overline{AC}$, lies inside of ABCD.

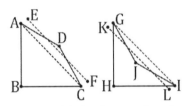

An **equiangular** polygon has congruent interior angles, but the sides are not necessarily congruent. An **equilateral** polygon has congruent sides, but not the interior angles are not necessarily congruent. A **regular** polygon has both congruent sides and congruent interior angles. A **regular** polygon is both **equilateral** and **equiangular**. An equilateral triangle is necessarily equiangular, but for a polygon with more than three sides it is possible for a polygon to be equilateral without being equiangular or for a polygon to be equiangular without being equilateral. For example, the hexagon below on the left is equiangular, but not equilateral (since the horizontal sides are longer than the others), whereas the hexagon below on the right is equilateral, but not equiangular (since the leftmost and rightmost interior angles are smaller than the others). Note that a regular polygon or an equiangular polygon must be convex, but an equilateral polygon can be concave or convex.

For any polygon (even if it is not regular) with N sides, the sum of the interior angles equals $180°(N - 2)$. Since 180° equates to π radians (Chapter 1), the sum of the interior angles in **radians** equals $\pi(N - 2)$. For example, the **sum** of the **interior** angles for a decagon equals $180°(10 - 2) = 180°(8) = 1440°$ (since $N = 10$ for a decagon), which equates to $\pi(10 - 2) = \pi(8) = 8\pi$ rad. For any polygon (even if it is not regular), the **sum** of the **exterior** angles equals $360°$ (regardless of the number of sides), which is equivalent to 2π rad.

For an **equiangular** polygon with N sides, **each interior** angle equals $180° - \frac{360°}{N}$, which is equivalent to $180°\left(1 - \frac{2}{N}\right)$, and **each exterior** angle equals $\frac{360°}{N}$. In **radians**, these formulas are $\pi - \frac{2\pi}{N}$ for each interior angle and $\frac{2\pi}{N}$ for each exterior angle. For example, each interior angle equals $180° - \frac{360°}{10} = 180° - 36° = 144°$ for an equiangular decagon and each exterior angle equals $\frac{360°}{10} = 36°$ (since $N = 10$ for a decagon). These equate to $\pi - \frac{2\pi}{10} = \frac{10\pi}{10} - \frac{2\pi}{10} = \frac{8\pi}{10} = \frac{4\pi}{5}$ rad and $\frac{2\pi}{10} = \frac{\pi}{5}$ rad.

The **apothem** is the distance from the center of a regular polygon to the midpoint of one of its sides. For example, in the diagram below, point C is the center of a regular heptagon. The center (C) is equidistant from the vertices and is also equidistant from the sides. Point M is the midpoint of side $\overline{AB}$. Line segment $\overline{CM}$ is an apothem.

A practical method of finding the **area** of a polygon is to split the polygon into triangles and quadrilaterals. For example, a regular hexagon can be divided into four equilateral triangles (left diagram below), and a regular octagon can be found by removing four right triangles from a square (right diagram below).

Chapter 10 Examples

Example 1. Determine the sum of the interior angles and the sum of the exterior angles for a nonagon in both degrees and radians.

A nonagon has 9 sides. Plug $N = 9$ into the formula for the sum of the interior angles for a polygon: $180°(9 - 2) = 180°(7) = 1260°$, which is equivalent to $\pi(9 - 2) = \pi(7)$ $= 7\pi$ rad. The sum of the exterior angles is $360°$, which is equivalent to 2π rad.

Example 2. Determine the angular measure of each interior angle and each exterior angle for a regular nonagon in both degrees and radians.

A nonagon has 9 sides. Plug $N = 9$ into the formulas for the interior and exterior angles of an equiangular polygon. Each interior angle equals $180° - \frac{360°}{9} = 180° - 40° = 140°$ and each exterior angle equals $\frac{360°}{9} = 40°$. In radians, these are $\pi - \frac{2\pi}{9} = \frac{9\pi}{9} - \frac{2\pi}{9} = \frac{7\pi}{9}$ rad and $\frac{2\pi}{9}$ rad.

Example 3. The two unknown angles marked θ are congruent in the diagram below. Determine θ.

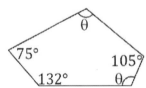

A pentagon has 5 sides. Plug $N = 5$ into the formula for the sum of the interior angles of a polygon: $180°(5 - 2) = 180°(3) = 540°$. The interior angles add up to $540°$.

$$75° + 132° + \theta + 105° + \theta = 540°$$
$$312° + 2\theta = 540°$$
$$2\theta = 540° - 312°$$
$$2\theta = 228°$$
$$\theta = \frac{228°}{2} = 114°$$

Example 4. Find the area, perimeter, and apothem of a regular hexagon that has an edge length equal to L.

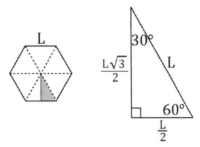

Divide the regular hexagon into six congruent triangles, as shown above on the left. A hexagon has N = 6 sides, such that each interior angle equals $180° - \frac{360°}{6} = 180° - 60°$ $= 120°$. Since the dashed lines above are angle bisectors, each congruent triangle is an equilateral triangle (60°-60°-60°). Recall from Chapter 6 that an equilateral triangle with edge length L has an area equal to $\frac{L^2\sqrt{3}}{4}$ (see Chapter 6). To find the area of the regular hexagon, multiply the area of the equilateral triangle by 6: $A = 6\frac{L^2\sqrt{3}}{4} = 3\frac{L^2\sqrt{3}}{2}$ (since 6/4 reduces to 3/2). The perimeter is P = 6L. The apothem equals the vertical side of the 30°-60°-90° half-triangle shown above on the right: $\frac{L\sqrt{3}}{2}$.

Example 5. Prove that the sum of the exterior angles for a polygon equals 360°.

Let ∠1, ∠2, ∠3, and so on, up to ∠N represent the interior angles. Let α, β, γ, and so on, up to η represent the exterior angles. Begin with the formula for the sum of the interior angles of a polygon.

$$∠1 + ∠2 + ∠3 + \cdots + ∠N = 180°(N - 2)$$

Distribute on the right-hand side. Recall from algebra that a(b − c) = ab − ac.

$$180°(N - 2) = 180°N - 180°(2) = 180°N - 360°$$

As shown above, each exterior angle is the supplement to one of the interior angles. For example, ∠1 + α = 180°, such that ∠1 = 180° − α. Replace each interior angle with 180° minus the exterior angle that is its supplement.

$$180° - α + 180° - β + 180° - γ + \cdots + 180° - η = 180°N - 360°$$

On the left-hand side, the value 180° appears N times (once for each exterior angle).

$$180°N - \alpha - \beta - \gamma - \cdots - \eta = 180°N - 360°$$

Subtract 180°N from both sides of the equation. It cancels out.

$$-\alpha - \beta - \gamma - \cdots - \eta = -360°$$

Multiply both sides of the equation by negative one. Every minus sign will change to a plus sign.

$$\alpha + \beta + \gamma + \cdots + \eta = 360°$$

Example 6. Show that each exterior angle equals $\frac{360°}{N}$ for an equiangular polygon with N sides.

As shown in Example 5, the sum of the exterior angles for any polygon equals 360°.

$$\alpha + \beta + \gamma + \cdots + \eta = 360°$$

For an equiangular polygon, the exterior angles are congruent. Set the N angles on the left-hand side equal. There are N exterior angles added together.

$$N\alpha = 360°$$

Divide by N on each side of the equation: $\alpha = \frac{360°}{N}$.

Chapter 10 Problems

Note: The diagrams are not drawn to scale.

1. Determine the sum of the interior angles and the sum of the exterior angles for each of the following in both degrees and radians.

(A) hexagon

(B) heptagon

(C) hendecagon

2. Determine the angular measure of each interior angle and each exterior angle for each of the following in both degrees and radians.

(A) regular pentagon

(B) regular octagon

(C) regular dodecagon

3. Is it possible for the angular measure of an interior angle to be the same value as the angular measure of an exterior angle for a regular polygon? If so, how many sides does the polygon have and what is the angular measure of each interior/exterior angle?

4. The sum of the interior angles of a polygon equals 2700°. Determine the number of sides that the polygon has.

5. Given only the sum of the exterior angles of a polygon, is it possible to determine how many sides the polygon has? Explain.

6. A regular polygon has interior angles equal to 156°. Determine the number of sides.

7. A regular polygon has exterior angles equal to 18°. Determine the number of sides.

8. Determine each of the three unknown interior angles in the diagram below.

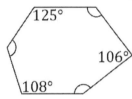

9. In the diagram below, the ratio of θ to φ is equal to 3:2. Determine θ and φ.

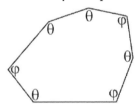

10. Determine the perimeter and area of the hexagon below.

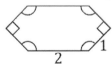

11. Determine d, α, the perimeter, and area of the pentagon below.

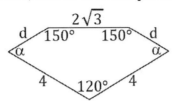

12. Find the perimeter, area, and apothem of a regular octagon that has an edge length equal to L.

13. The equilateral concave decagon below has five pairs of collinear sides. The dashed lines on the right form a regular pentagon. Find interior angles θ and φ.

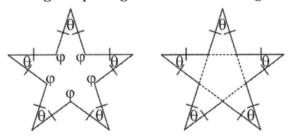

14. Find d, ∠1, the perimeter, and the area of the concave kite below.

15. The diagram below shows an **<u>incomplete</u>** convex polygon. The interior angles start at $\angle 1$, $\angle 2$, $\angle 3$, and continue to the last interior angle, $\angle N$. The exact number of interior numbers (and sides) is not known. Point P lies inside of the polygon. The dashed lines split the polygon into N triangles. Prove that the interior angles add up to $180°(N - 2)$. Note: Other solid sides that complete the polygon have not been drawn.

16. Show that each interior angle equals $180° - \frac{360°}{N}$ for an equiangular polygon with N sides.

17. Consider the statement, "If a polygon is not equiangular, it is not regular."
(A) Is this statement true?

(B) What is the converse of the given statement? Is the converse true?

(C) What is the inverse of the given statement? Is the inverse true?

(D) What is the contrapositive of the given statement? Is the contrapositive true?

Answer Key

Chapter 1 Answers

1. $\angle CED = 34°$. There are two pairs of complements: $\angle AEB$ & $\angle BEC$ and $\angle BEC$ & $\angle CED$.

2. $\angle 1 = 33°$, $\angle 2 = 147°$, and $\angle 3 = 123°$. Check: $33° + 147° + 123° + 57° = 360°$.

3. $\theta = \frac{\pi}{3}$ rad $= 60°$, $\varphi = \frac{\pi}{6}$ rad $= 30°$, and $\psi = \frac{5\pi}{6}$ rad $= 150°$.
Check: $60° + 30° + 150° + 30° + 90° = 360°$.

4. $\angle AEB = \angle BEC = \angle CED = 30° = \frac{\pi}{6}$ rad. Check: $30° + 30° + 30° = 90°$.

5. $\theta = 36°$. Check: $36° + 36° + 36° + 36° + 36° = 5 \times 36° = 180°$.

6. $\varphi = 60° = \frac{\pi}{3}$ rad. Check: $60° + 60° + 60° + 60° + 60° + 60° = 6 \times 60° = 360°$.

7. $\angle AEB = 90° = \frac{\pi}{2}$ rad, $\angle BEC = 30° = \frac{\pi}{6}$ rad, and $\angle CED = \angle DEA = 120° = \frac{2\pi}{3}$ rad.
Check: $90° + 30° + 120° + 120° = 360°$.

8. $\angle 1 = 40°$, $\angle 2 = 54°$, and $\angle 3 = 40°$. Check: $40° + 140° + 40° + 54° + 86° = 360°$.

9. $\theta = \frac{4\pi}{3}$ rad $= 240°$. Check: $120° + 240° = 360°$.

10. $\angle ABC = 60°$ and $\angle CBD = 30°$. Check: $60° + 30° = 90°$.

11. $\angle 1 = 100°$, $\angle 2 = 60°$, and $\angle 3 = 20°$. Check: $100° + 60° + 20° = 180°$.

12. $\psi = 15° = \frac{\pi}{12}$ rad. Check: $5 \times 15° + 4 \times 15° + 15 \times 15° = 75° + 60° + 225° = 360°$.

13. An angle with a single arc is $36°$. An angle with a double arc is $72°$. Check:
$36° + 36° + 72° + 36° + 72° + 36° + 72° = 4 \times 36° + 3 \times 72° = 144° + 216° = 360°$.

14. AD $= 7.2$ units. Check: $2.4 + 2.4 + 2.4 = 7.2$.

15. PU $= 22.5$ units. Check: $4.5 + 4.5 + 4.5 + 4.5 + 4.5 = 5 \times 4.5 = 22.5$.

16. AB $= 4$ units, CF $= 12$ units, and BH $= 24$ units. Note: Although there are 10 points, there are 9 gaps between the points: $36 \div 9 = 4$.

17. GH $= 2$ units and HI $= 4$ units. Check: $2 + 4 + 4 + 2 = 12$.

18. XY $= 0.375$ units (equivalent to 3/8 units). Check: $1.5 + 0.75 + 0.375 + 0.375 = 3$.

19. $\angle FEG = 54°$, $\angle DEH = 54°$, and $\angle GEH = 126°$. Notes: $\angle ADE$ and $\angle DEH$ are alternate interior angles. $\angle ADE$ and $\angle FEG$ are corresponding angles.

20. $\angle 1 = 117°$, $\angle 2 = 63°$, $\angle 3 = 63°$, and $\angle 4 = 117°$. Notes: $\angle 2$ and $\angle 3$ are alternate interior angles. $\angle 1$ and $\angle 4$ are corresponding angles.

21. Left diagram: the left and right lines intersect above the transversal.

Why? Because $28° + 148° = 176°$ is less than $180°$. ($148°$ is the supplement to $32°$.)

Middle diagram: the left and right lines intersect below the transversal.

Why? Because $31° + 151° = 182°$ is greater than $180°$.

Right diagram: the left and right lines are parallel. Why? Because $27° + 153° = 180°$.

22. $\angle VWX = 67°$. Check: $72° + 67° + 41° = 180°$.

23. $\theta = \frac{5\pi}{6}$ rad $= 150°$. Notes: The trick is to draw another line, such that the new line segment is parallel to $\overline{AB}$ and passes through C. In the diagram below, $\overline{FC} \parallel \overline{AB} \parallel \overline{DE}$. It follows that $\angle ABC + \angle BCF = 180°$ and $\angle FCD + \theta = 180°$. Check: $120° + 60° = 180°$, $60° + 30° = 90°$, and $30° + 150° = 180°$.

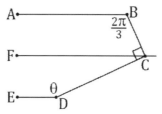

24. $\angle LMN = 143°$. Notes: The trick is to draw a new line segment parallel to $\overline{KL}$ that passes through M. In the diagram below, $\overline{JM} \parallel \overline{KL}$. It follows that $\angle KLM + \angle LMJ = 180°$ and $\angle LMN = \angle LMJ + 90°$. Check: $127° + 53° = 180°$.

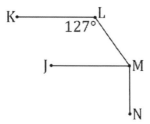

25. $\angle 1 = 141°$ and $\angle 2 = 141°$. Notes: $\angle QRS$, $\angle 1$, and $\angle 2$ are alternate interior angles. To see this, extend each line segment, as shown below.

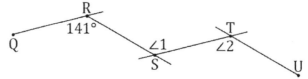

26. If the two angles are congruent, they will both be right angles. Otherwise, there will be one acute angle and one obtuse angle.

27. There can be at most one obtuse angle, and if there is one obtuse angle there will also be two acute angles. There can be at most one right angle, and if there is one right angle there will also be two acute angles. If there are no right or obtuse angles, there will be three acute angles.

28. Main ideas: $\angle 1$ and $\angle 2$ are supplements. $\angle 2$ and $\angle 4$ add up to $180°$ because lines $\overleftrightarrow{AB}$ and $\overleftrightarrow{CD}$ are parallel. Write down an equation for each step and set the equations equal to one another.

29. Main ideas: $\angle 2$ and $\angle 3$ are vertical angles. $\angle 3$ and $\angle 6$ are alternate interior angles.

30. (A) Yes. (B) "If two lines are not parallel, they are perpendicular." The converse is false. Two lines can intersect without being perpendicular.

(C) "If two lines are not perpendicular, they are parallel." The inverse is false.

(D) "If two lines are parallel, they are not perpendicular." The contrapositive is true.

Chapter 2 Answers

1. $\theta = 19°$. Check: $19° + 127° + 34° = 180°$.

2. $\angle BCA = 32°$. Note: $\angle ABC$ is a right angle. Check: $58° + 90° + 32° = 180°$.

3. $\angle 2 = \frac{7\pi}{24}$ rad $= 52.5°$. Check: $\frac{3\pi}{8} + \frac{7\pi}{24} + \frac{\pi}{3} = \frac{9\pi}{24} + \frac{7\pi}{24} + \frac{8\pi}{24} = \frac{24\pi}{24} = \pi$.
Check in degrees: $67.5° + 52.5° + 60° = 180°$.

4. $\varphi = 126°$. The tick marks indicate that the triangle is isosceles. The angles opposite to the congruent sides are equal. Check: $126° + 27° + 27° = 180°$.

5. $\angle 1 = \frac{3\pi}{10}$ rad $= 54°$. The tick marks indicate that the triangle is isosceles. The angles opposite to the congruent sides are equal. Check: $\frac{3\pi}{10} + \frac{3\pi}{10} + \frac{2\pi}{5} = \frac{3\pi}{10} + \frac{3\pi}{10} + \frac{4\pi}{10} = \frac{10\pi}{10} = \pi$.
Check in degrees: $54° + 54° + 72° = 180°$.

6. $\angle ABD = 118°$. Note: $\angle ABD$ is an exterior angle. Check: $42° + 76° = 118°$.

7. $\alpha = 59°$. Note: α is an exterior angle. Check: $28° + 31° = 59°$.

8. $\angle 4 = \frac{3\pi}{4}$ rad $= 135°$. Notes: $\frac{2\pi}{3}$ rad is an exterior angle. The corresponding interior angle is its supplement: $\pi - \frac{2\pi}{3} = \frac{3\pi}{3} - \frac{2\pi}{3} = \frac{\pi}{3}$ rad. The two interior angles, $\frac{5\pi}{12}$ rad and $\frac{\pi}{3}$ add up to the exterior angle at the third vertex. Check: $\frac{5\pi}{12} + \frac{\pi}{3} = \frac{5\pi}{12} + \frac{4\pi}{12} = \frac{9\pi}{12} = \frac{3\pi}{4}$.
Check in degrees: $75° + 60° = 135°$.

9. $\tau = \frac{2\pi}{3}$ rad $= 120°$. Notes: This is an equilateral triangle. Each interior angle is $60°$ and each exterior angle is $120°$. Check: $\frac{\pi}{3} + \frac{\pi}{3} = \frac{2\pi}{3}$. Check in degrees: $60° + 60° = 120°$.

10. $\angle BEF = 84°$. Note: The given angles and $\angle BEF$ are all exterior angles. Since each exterior angle is at a different vertex, the sum of these angles is $360°$.
Check: $129° + 147° + 84° = 360°$.

11. $\angle 1 = 114°, \angle 2 = 34°, \angle 3 = 32°, \angle 4 = 66°, \angle 5 = 146°, \angle 6 = 34°, \angle 7 = 146°, \angle 8 = 148°, \angle 9 = 148°$ and $\angle 10 = 66°$. Notes: $\angle 1$ & $114°$, $\angle 4$ & $\angle 10$, $\angle 2$ & $\angle 6$, $\angle 5$ & $\angle 7$, $\angle 3$ & $32°$, and $\angle 8$ & $\angle 9$ are vertical angles; $\angle 1, \angle 2$, and $\angle 3$ are interior angles; and $\angle 4, \angle 10, \angle 5, \angle 7, \angle 8$, and $\angle 9$ are exterior angles. Interior angle check: $114° + 34° + 32° = 180°$. Exterior angle check: $66° + 146° + 148° = 360°$. Vertical angle check: $2 \times 114° + 2 \times 66° = 360°$, $2 \times 34° + 2 \times 146° = 360°$, and $2 \times 32° + 2 \times 148° = 360°$.

12. $\angle ABD = 58°, \angle DBE = 79°,$ and $\angle BED = 43°.$ Note: $\angle ABD$ & $\angle BDE$ and $\angle CBE$ & $\angle BED$ are alternate interior angles (Chapter 1). Check: $58° + 79° + 43° = 180°.$

13. $\theta = 54°, \varphi = 36°,$ and $\chi = 54°.$ Note: φ and χ are complementary angles. Checks: $54° + 36° + 90° = 180°$ for all three triangles (including the outside triangle).

14. $\angle 1 = \angle 2 = 61°, \angle 3 = 119°,$ and $\angle 4 = 29°.$ Notes: $\angle 1$ and $\angle 2$ are equal because the angles opposite to the congruent sides of an isosceles triangle are equal. Angle $\angle 3$ is an exterior angle of the left triangle. Checks: $61° + 61° + 58° = 180°, 119° + 29° + 32° = 180°, 61° + 32° + (58° + 29°) = 180°$ (for the surrounding triangle).

15. $\alpha = 14°, \beta = 23°, \gamma = 119°,$ and $\delta = 114°.$ Note: $\delta, \gamma,$ and $127°$ add up to a full circle $(360°).$ Checks: $35° + 26° + 119° = 180°, 14° + 39° + 127° = 180°, 23° + 43° + 114° = 180°,$ and $(43° + 35°) + (14° + 26°) + (23° + 39°) = 180°$ (for the big triangle).

16. Left: acute, equilateral, and equiangular are the three most precise terms.

Middle: obtuse, scalene. Note: The top interior angle is $99°.$

Right: right, isosceles. Since two interior angles are marked as congruent, the triangle is isosceles. Since two interior angles are $\frac{\pi}{4}$ rad $= 45°,$ the third angle is $\frac{\pi}{2}$ rad $= 90°.$

17. Main ideas: An exterior angle and interior angle at the same vertex are supplements. The three interior angles add up to $180°.$ Write down an equation for each idea and set the equations equal to one another.

18. Main idea: $\angle 4$ & $\angle 6$ are vertical angles below.

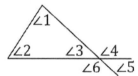

19. Main ideas: The three interior angles add up to $180°.$ At one vertex, the interior and exterior angles are supplements. Rewrite the interior angle sum in terms of exterior angles.

20. Main ideas: The interior angles are congruent and add up to $180°.$ The exterior angles are congruent and add up to $360°.$

21. (A) Yes. (B) "If the smallest angles are complements, the triangle is right." True.

(C) "If the triangle is not right, the smallest angles are not complements." True.

(D) "If the smallest angles are not complements, the triangle is not right." True.

Chapter 3 Answers

1. ∠ACB and ∠DCE are vertical angles, ∠CAB and ∠CED are alternate interior angles, and $\overline{AB}$ and $\overline{DE}$ are marked as congruent (AAS). Note: Since ∠ABC and ∠EDC are also alternate interior angles, it is possible to prove this using ASA.

2. FH = HI, ∠FHG and ∠IHJ are vertical angles, and GH = HJ (SAS). Notes: A formal proof would give statements using congruence, like $\overline{FH} \cong \overline{HI}$. Formal geometry proofs often use two columns: one column for mathematical statements and a second column for the corresponding reasons. Unlike Problem 1, this problem does not state whether or not any of the line segments are parallel.

3. Two pairs of sides are marked as congruent and the third side is shared (SSS). Notes: In a formal proof, the statement for the third side would be something like $\overline{AB} \cong \overline{AB}$ and the reason would be the reflexive property (Chapter 1). Do not assume that any of the line segments are parallel.

4. ∠LMK and ∠NMO are vertical angles, ∠LKM and ∠NOM are marked as congruent, and $\overline{KL}$ and $\overline{NO}$ are marked as congruent (AAS).

5. PS = QS because ΔPQS is isosceles, ∠PSR and ∠QSR are marked as congruent, and side RS is shared by both triangles (SAS).

6. WZ = YZ, ∠WZX and ∠XZY are both right angles, and XZ is shared (SAS).

7. ∠AEB and ∠BDC are both right angles, ∠BAE and ∠CBD are congruent (since ∠BAE and ∠ABE are complementary angles, and since ∠ABE and ∠CBD are complementary angles), and AB = BC because ΔABC is isosceles (AAS). Notes: DBE is a straight angle, such that ∠ABE, ∠ABC, and ∠CBD add up to 180°. Since ∠ABC is a right angle, it follows that ∠ABE and ∠CBD are complementary angles. Do not assume that any of the line segments are parallel.

8. PS = QR, PT = QT because ΔPQT is isosceles, and ST = RT because ΔRST is isosceles (SSS). Note: Do not assume that any of the line segments are parallel.

9. ∠FHG and ∠IHJ are vertical angles, GH = HJ because ΔGHJ is isosceles, and ∠FGH and ∠HJI are congruent since they are complements to congruent angles (ASA). Notes: Do not assume that any of the line segments are parallel (but if you first prove that $\overline{FG} \parallel \overline{IJ}$, then you may use this). ∠FGH = 90° − ∠HGJ and ∠HJI = 90° − ∠GJH.

10. KN = LN because ΔKLN is isosceles, ∠KNO and ∠MLN are both right angles, and NO = ML are marked as congruent (SAS).

11. ∠ACD and ∠BDE are corresponding angles, CD = DE, and ∠ADC and ∠BED are corresponding angles (ASA).

12. ∠1 is opposite to b, ∠2 is opposite to a, and ∠3 is opposite to c.

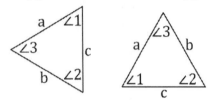

13. α is opposite to j, θ is opposite to x, and φ is opposite to q. Also note that θ is formed by j and q, α is formed by q and x, and φ is formed by j and x.

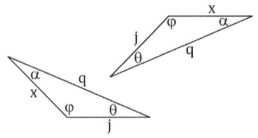

14. (A) Yes. (B) "If all three interior angles are congruent, two triangles are congruent." False. As Chapter 4 will show, two triangles with three congruent interior angles are similar, but may not be congruent.

(C) "If two triangles are not congruent, not all three interior angles are congruent." False. As Chapter 4 will show, two triangles that are similar (but not congruent) have congruent interior angles.

(D) "If not all three interior angles are congruent, two triangles are not congruent." True. The contrapositive is always true if the original statement is true. In this case, if two triangles are congruent, all three corresponding sides and all three corresponding interior angles are congruent. Thus, if all three interior angles are not congruent, the two triangles are not congruent.

Chapter 4 Answers

1. $\angle ACB$ and $\angle DCE$ are vertical angles, and $\angle CAB$ and $\angle CDE$ are alternate interior angles (AA). BC = 21. Check: AC/BC = CD/CE gives 12/21 = 16/28 = 4/7. Notes: AC and CD are opposite to $\angle ABC$ and $\angle CED$, while BC and CE are opposite to $\angle BAC$ and $\angle CDE$.

2. $\angle VXW$ and $\angle YXZ$ are vertical angles, and VX/WX = XZ/XY since 9/18 = 11/22 = 1/2 (SAS similarity). $\angle VWX = 30°$ and $\angle WVX = 90°$. Notes: $\angle XZY = 90°$ because $\angle XZY$, $\angle YXZ = 60°$, and $\angle XYZ = 30°$ add up to 180°. $\angle VWX = \angle XYZ$ because these angles are opposite to VX and XZ. $\angle WVX = \angle XZY$ because these angles are opposite to WX and XY.

3. $\angle HFI$ is shared by both triangles (a formal proof would give the reflexive property as the reason that $\angle GFJ = \angle HFI$), and GF/FJ = HF/FI since 30/39 = 40/52 = 10/13 (SAS similarity). HI = 32. Check: GJ/FG = HI/FH gives 24/30 = 32/40 = 4/5. Notes: 10 and 13 are not sides of a triangle. Coincidentally, 10/13 is what 30/39 and 40/52 reduce to. FH = 30 + 10 = 40. FI = 39 + 13 = 52.

4. $\angle RSU \cong \angle STU$ because $\angle RSU$ and $\angle STU$ are each complements to $\angle TSU$, and $\angle SRU$ and $\angle SUT$ are both right angles (AA). ST = 289/8 = 36.125 and TU = 255/8 = 31.875. Checks: RU/RS = SU/TU gives 8/15 = 17/(255/8), and RS/SU = TU/ST gives 15/17 = 255/8÷(289/8). Notes: 17÷(255/8) = 17×(8/255) = 136/255 = 8/15 if you divide 136 and 255 each by 17, and 255/8÷(289/8) = (255/8)×(8/289) = 255/289 = 15/17 if you divide 255 and 289 each by 17.

5. MN:KN:KM = 24:36:48 is equivalent to LM:KM:KL = 32:48:64 because both triple ratios reduce to 2:3:4 (SSS similarity). $\angle MKN \cong \angle LKM$, $\angle KMN \cong \angle KLM$, and $\angle KNM \cong \angle KML$. Notes: Divide 24, 36, and 48 each by 12 and divide 32, 48, and 64 each by 16 to see that each·triple ratio reduces to 2:3:4. Alternatively, note that 24/36 = 32/48 = 2/3 and 36/48 = 48/64 = 3/4. $\angle MKN$ and $\angle LKM$ are opposite to the short sides (24 and 32), $\angle KMN$ and $\angle KLM$ are opposite to the middle sides (36 and 48), and $\angle KNM$ and $\angle KML$ are opposite to the long sides (48 and 64).

6. $\angle CBD$ and $\angle FAE$ are corresponding angles, and $\angle CDB$ and $\angle FEA$ are corresponding angles (AA). AF = 22.5 and DE = 7. Checks: CD/BC = FE/FA gives 8/9 = 20/22.5, and CD/BD = FE/AE gives 8/10 = 20/25 = 4/5. Notes: AE = 8 + 10 + 7 = 25. To see that 20/22.5 = 8/9, multiply 20 and 22.5 each by 2 to get 40/45, then divide 40 and 45 by 5.

7. ∠PQS ≅ ∠QRS because ∠PQS and ∠QRS are each complements to ∠RQS, and ∠QPS ≅ ∠RQS because ∠QPS and ∠RQS are complements to ∠PQS and ∠QRS (AA). QR = 40 and RS = 32. Checks: PS:QS:PQ = 18:24:30, QS:RS:QR = 24:32:40, and PQ:QR:PR = 30:40:50 each reduce to 3:4:5. Note: PR = PS + RS = 18 + 32 = 50.

8. ∠XZY and ∠VWY are marked as congruent, and XZ/YZ = VW/WY since 35/45 = 70/90 = 7/9 (SAS similarity). VZ = 15. Check: XY/ZY = VY/WY gives 30/45 = 60/90 = 2/3. Notes: WY = 30 + 60 = 90. VY = 45 + 15 = 60. XZ is not perpendicular to WY.

9. ∠UTV and ∠WVX are alternate interior angles, and ∠TVU and ∠VXW are alternate interior angles (AA). TU = 100 and UV = 105. Checks: TU/UV = WV/WX gives 100/105 = 20/21, and TU/TV = WV/VX gives 100/120 = 20/24 = 5/6. Notes: UV and WX are opposite to ∠UTV and ∠WVX, TU and WV are opposite to ∠TVU and ∠VXW, and TV and VX are opposite to ∠TUV and ∠VWX. TV = 96 + 24 = 120. One way to see that 100/105 reduces to 20/21 is to divide 100 and 105 each by 5.

10. ∠BAD ≅ ∠BDA because ΔBAD is isosceles, and ∠BAC ≅ ∠ABC because ΔABC is isosceles (AA). BC = 25 and CD = 9. Check: AD:AB:BD = 16:20:20 and AB:AC:BC = 20:25:25 each reduce to 4:5:5. Divide 16 and 20 each by 4, and divide 20 and 25 each by 5 to see this. Notes: AC = 16 + 9 = 25. ∠ABD ≅ ∠ACB as these angles are opposite to the short sides (AD = 16 and AB = 20). ∠BAD ≅ ∠BAC are the same angle.

11. ΔKMO~ΔLNO. ∠KOM ≅ ∠LNO, ∠MKO ≅ ∠LON, and ∠KMO ≅ ∠NLO. KM:MO:KO corresponds to LO:LN:NO. Check: KM:MO:KO = 81:90:99 and LO:LN:NO = 90:100:110 each reduce to 9:10:11. Divide 81, 90, and 99 each by 9, and divide 90, 100, and 110 each by 10 to see that each triple ratio reduces to 9:10:11. Notes: Short sides KM and LO are opposite to ∠KOM and ∠LNO, medium sides MO and LN are opposite to ∠MKO and ∠LON, and long sides KO and NO are opposite to ∠KMO and ∠NLO. Side MO = 90 is the medium length of ΔKMO, whereas side LO = 90 is the short length of ΔLNO.

12. (A) Yes. The first part is AA. The second part is SSS similarity. If one is true, so is the other. (B) "If the sides of the triangles come in the same proportions, two of the three interior angles are congruent." True. In fact, all three interior angles are congruent. (C) "If two interior angles are not both congruent, the sides do not come in the same proportions." True. One angle that is not congruent is enough. (D) "If the sides do not come in the same proportions, two interior angles are not both congruent." True.

Chapter 5 Answers

1. $c = 13$. Check: $5^2 + 12^2 = 25 + 144 = 169 = 13^2$.

2. $a = 15$. Check: $8^2 + 15^2 = 64 + 225 = 289 = 17^2$. Note: The hypotenuse is 17.

3. $d = 5\sqrt{5}$. Check: $5^2 + 10^2 = 25 + 100 = 125 = \left(5\sqrt{5}\right)^2$. Notes: $\sqrt{125}$ is equivalent to $5\sqrt{5}$ because $\sqrt{125} = \sqrt{25(5)} = \sqrt{25}\sqrt{5} = 5\sqrt{5}$. The preferred form is $5\sqrt{5}$.

4. $e = 6\sqrt{2}$. Check: $3^2 + \left(6\sqrt{2}\right)^2 = 9 + (6)^2\left(\sqrt{2}\right)^2 = 9 + 36(2) = 9 + 72 = 81 = 9^2$. Notes: $\sqrt{72}$ is equivalent to $6\sqrt{2}$ because $\sqrt{72} = \sqrt{36(2)} = \sqrt{36}\sqrt{2} = 6\sqrt{2}$. The preferred form is $6\sqrt{2}$. The hypotenuse is 9.

5. $\theta = 45°$ and $h = 5\sqrt{2}$. Notes: The three interior angles add up to 180°. This is a 45°-45°-90° triangle. $h = 5\sqrt{2}$ is $\sqrt{2}$ times larger than 5.

6. $\angle 2 = 60°$, $b = \frac{\sqrt{3}}{2}$, and $c = 1$. Notes: The three interior angles add up to 180°. This is a 30°-60°-90° triangle. $c = 1$ is twice $\frac{1}{2}$ and $b = \frac{\sqrt{3}}{2}$ is $\sqrt{3}$ times $\frac{1}{2}$.

7. $\varphi = \frac{\pi}{6}$ rad (equivalent to 30°), $e = 6$, and $a = 12$. Notes: $\frac{\pi}{3}$ rad is equivalent to 60°. The three interior angles add up to 180°. This is a 30°-60°-90° triangle. $6\sqrt{3}$ is $\sqrt{3}$ times larger than $e = 6$ and $a = 12$ is twice $e = 6$.

8. $\angle 1 = 45°$ and $d = 2$. Notes: The two acute angles are marked as congruent, which makes this triangle isosceles. The three interior angles add up to 180°. This is a 45°-45°-90° triangle. $d = 2$ is $\sqrt{2}$ larger than $\sqrt{2}$ because $\sqrt{2}\sqrt{2} = 2$. The ratio $\sqrt{2}:\sqrt{2}:2$ is equivalent to the ratio $1:1:\sqrt{2}$ (since $1\sqrt{2} = \sqrt{2}$ and $\sqrt{2}\sqrt{2} = 2$).

9. $\angle 4 = 30°$, $t = 4$, and $w = 4\sqrt{3}$. Notes: The three interior angles add up to 180°. This is a 30°-60°-90° triangle. 8 is twice $t = 4$ and $w = 4\sqrt{3}$ is $\sqrt{3}$ times $t = 4$.

10. $\alpha = \frac{\pi}{3}$ rad (equivalent to 60°), $n = \sqrt{6}$, and $m = 2\sqrt{2}$. Notes: $\frac{\pi}{6}$ rad is equivalent to 30°. The three interior angles add up to 180°. This is a 30°-60°-90° triangle. $n = \sqrt{6}$ is $\sqrt{3}$ times larger than $\sqrt{2}$ and $m = 2\sqrt{2}$ is twice $\sqrt{2}$.

11. $\theta = 45°$ and $k = \sqrt{3}$. Notes: The two legs are marked as congruent, which makes this triangle isosceles. The three interior angles add up to 180°. This is a 45°-45°-90° triangle. $\sqrt{6}$ is $\sqrt{2}$ larger than $k = \sqrt{3}$ because $\sqrt{3}\sqrt{2} = \sqrt{6}$.

12. Main ideas: Draw a vertical line (like the dashed line below) from the top vertex to the base in order to divide the 30°-30°-120° triangle into two 30°-60°-90° triangles. Let $c = \frac{1}{2}$ in the diagram below. Then $a = b = \frac{\sqrt{3}}{2}$ and the hypotenuse of each triangle equals 1. Note that $a = b = \frac{\sqrt{3}}{2}$ is $\sqrt{3}$ times $c = \frac{1}{2}$ and that 1 is twice $c = \frac{1}{2}$. The sides of the 30°-30°-120° are 1, 1, and $a + b = \frac{\sqrt{3}}{2} + \frac{\sqrt{3}}{2} = \frac{2\sqrt{3}}{2} = \sqrt{3}$, which make the ratio $1:1:\sqrt{3}$.

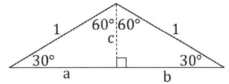

13. Main ideas: One leg (11) is congruent and the hypotenuse (14) is shared by the two triangles. The triangles are congruent according to LH. $p = q = 5\sqrt{3}$. Check: $11^2 + \left(5\sqrt{3}\right)^2$ $= 121 + 5^2\left(\sqrt{3}\right)^2 = 121 + 25(3) = 121 + 75 = 196 = 14^2$. Since 14 is opposite to the right angle, 14 is the hypotenuse. The hypotenuse is also the longest side.

14. Main ideas: The legs come in the same ratio: $CD/BC = 9/12 = 3/4$ and $BC/AB = 12/16 = 3/4$. $\triangle ABC \sim \triangle BCD$ according to SAS similarity. $AC = 20$ and $BD = 15$. Checks: $BC/AC = 12/20 = 3/5$ agrees with $CD/BD = 9/15 = 3/5$. The Pythagorean theorem gives $9^2 + 12^2 = 81 + 144 = 225 = 15^2$ and $12^2 + 16^2 = 144 + 256 = 400 = 20^2$.

15. Main ideas: The three interior angles of any triangle add up to 180°. Plug 90° in for one of the interior angles. Subtract 90° from both sides.

16. Main ideas: The top and bottom sides of the square are parallel, and the right and left sides of the square are parallel. Find corresponding angles in each triangle. The triangles are similar according to AA. The sides of the three similar triangles come in the same proportions: $d{:}L{:}f = L{:}e{:}g = a{:}b{:}c$. Since $d{:}L{:}f = a{:}b{:}c$, it follows that

$$\frac{d}{L} = \frac{a}{b}$$

(using the first part of d:L:f and the first part of a:b:c). Note that $d + L = a$, such that $d = a - L$. Plug $d = a - L$ into $d/L = a/b$ to get:

$$\frac{a - L}{L} = \frac{a}{b}$$

Cross multiply. Multiply both sides of the equation by bL.

$$(a - L)b = aL$$

Distribute: $(a - L)b = ab - Lb$.

$$ab - Lb = aL$$

Divide every term by abL.

$$\frac{ab}{abL} - \frac{bL}{abL} = \frac{aL}{abL}$$

Note that ab cancels in the first term, bL cancels in the second term, and aL cancels on the right.

$$\frac{1}{L} - \frac{1}{a} = \frac{1}{b}$$

Add 1/a to both sides of the equation.

$$\frac{1}{L} = \frac{1}{a} + \frac{1}{b}$$

17. (A) False. A triangle that has exactly two acute angles could be obtuse or right.

(B) "If a triangle is a right triangle, it has exactly two acute interior angles." True.

(C) "If does not have exactly two acute interior angles, the triangle is not right." True. Note that if the converse of a statement is true, the inverse is also true.

(D) "If a triangle is not a right triangle, it does not have exactly two acute interior angles." False. Since the original statement is false, the contrapositive is also false. Note that the contrapositive is only true if the original statement is true.

Chapter 6 Answers

1. P = 12 and A = 6. Notes: h = 3 such that $h^2 + 4^2 = 3^2 + 4^2 = 9 + 16 = 25 = 5^2$. The base is b = 4. The area is bh/2.

2. A = 6. Notes: P/2 = 6. The three sides are 3, 4, and 5.
$$A = \sqrt{6(6-3)(6-4)(6-5)} = \sqrt{6(3)(2)(1)} = \sqrt{6(6)} = 6$$

3. P = 42 and A = 84. Notes: d = 9 such that $d^2 + 12^2 = 9^2 + 12^2 = 81 + 144 = 225$ $= 15^2$ and c = 13 such that $5^2 + 12^2 = 25 + 144 = 169 = 13^2 = c^2$. The base is 14. The area is bh/2.

4. A = 84. Notes: P/2 = 21. The three sides are 13, 14, and 15.
$$A = \sqrt{21(21-13)(21-14)(21-15)} = \sqrt{21(8)(7)(6)}$$
$$= \sqrt{(3 \times 7)(4 \times 2)(7)(3 \times 2)} = \sqrt{7^2 4^2 3^2} = 7 \times 4 \times 3 = 7 \times 12 = 84$$

5. P = 36 and A = 36. Notes: p = 6 such that $p^2 + 8^2 = 6^2 + 8^2 = 36 + 64 = 100 = 10^2$ and q = 9 such that the p + q = 15 and $(p+q)^2 + 8^2 = 15^2 + 8^2 = 225 + 64 = 289 = 17^2$. The base of the obtuse triangle is b = 9. The area is bh/2 = (9)(8)/2 = 36.

6. A = 36. Notes: P/2 = 18. The three sides of the obtuse triangle are 9, 10, and 17.
$$A = \sqrt{18(18-9)(18-10)(18-17)} = \sqrt{18(9)(8)(1)}$$
$$= \sqrt{(9 \times 2)(9)(4 \times 2)} = \sqrt{9^2 4^2} = 9 \times 4 = 36$$

7. A = 14. Notes: b = 7, h = 4, and bh/2 = 7(4)/2 = 28/2 = 14.

8. All three triangles have the same area. Notes: All three triangles have the same base since $BC \cong DF \cong GH$. All three triangles have the same height since $\overline{AI} \parallel \overline{BH}$. Since the base and height are the same for each, it follows that bh/2 is the same for each.

9. P = 24 and A = $16\sqrt{3}$. Notes: The area of an equilateral triangle with edge length d is $\frac{d^2\sqrt{3}}{4}$. The base is 8 and the height is $4\sqrt{3}$. (Cut the equilateral triangle in half to make two 30°-60°-90° triangles with sides 4, $4\sqrt{3}$, and 8 to see that the height is $4\sqrt{3}$.)

10. P = $12 + 8\sqrt{3}$, which is equivalent to $= 4(3 + 2\sqrt{3})$, and A = $12\sqrt{3}$. Notes: Draw the height as shown on the next page to divide the triangle into two congruent 30°-60°-90° triangles. The base is b = 12 and the height is h = $2\sqrt{3}$ (which is equivalent to $\frac{6}{\sqrt{3}}$ since $\frac{6}{\sqrt{3}}\frac{\sqrt{3}}{\sqrt{3}} = \frac{6\sqrt{3}}{3} = 2\sqrt{3}$). Note that c = $4\sqrt{3}$. The ratio h:6:c = $2\sqrt{3}$:6:$4\sqrt{3}$ reduces to 1:$\sqrt{3}$:2 (when each number is divided by $2\sqrt{3}$).

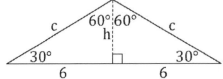

$P = b + c + c = 12 + 4\sqrt{3} + 4\sqrt{3} = 12 + 8\sqrt{3} = 4\left(3 + 2\sqrt{3}\right)$ and $A = \frac{bh}{2} = \frac{(12)(2\sqrt{3})}{2} =$ $12\sqrt{3}$. Either answer, $12 + 8\sqrt{3}$ or $4\left(3 + 2\sqrt{3}\right)$, is fine for the perimeter.

11. $A = 42$. Notes: $\triangle ABD$ and $\triangle ABE$ have the same height (AC), but the base of $\triangle ABE$ is 3/2 times the base of $\triangle ABD$. Since $A = \frac{1}{2}bh$ and h is the same, the ratio of the areas equals the ratio of the bases. Multiply 28 by 3/2 to get $28 \times 3/2 = 84/2 = 42$. If two triangles have the same height or the same base, the ratio of the areas equals the ratio of the bases or heights. For example, the ratio of the area of $\triangle ABD$ to the area of $\triangle ABE$ equals BD/BE because they have the same height (AC).

12. $A = 108$. Notes: Since all three interior angles are congruent, the two triangles are similar according to AA. The base and height (and all three sides) come in the same proportion. The base and height of $\triangle DEF$ are each 3 times the base and height of $\triangle ABC$. Since $A = \frac{1}{2}bh$, if the base is tripled and the height is also tripled, the area increases by a factor of 9. The area of $\triangle DEF$ equals $12 \times 9 = 108$. Alternatively, use Heron's formula. If two triangles are similar, the ratio of the areas equals the square of the ratio of any corresponding sides. For example, the ratio of the area of $\triangle DEF$ to the area of $\triangle ABC$ is $(DE/AB)^2$, which is the same as $(DF/AC)^2$ or $(EF/BC)^2$.

13. Main ideas: This is similar to Example 5, except for subtracting the areas of right triangles instead of adding them. The area of $\triangle WXZ$ is ah/2. The base of $\triangle WYZ$ equals $a + c$. The area of $\triangle WYZ$ is $(a + c)h/2$. The area of $\triangle XYZ$ can be found by subtraction:
$$\frac{(a + c)h}{2} - \frac{ah}{2} = \frac{ah + ch}{2} - \frac{ah}{2} = \frac{ah + ch - ah}{2} = \frac{ch}{2}$$

14. (A) $d^2 + h^2 = a^2$ and $(b - d)^2 + h^2 = c^2$. The first equation leads to $h^2 = a^2 - d^2$. Subtract $(b - d)^2 + h^2 = c^2$ from $d^2 + h^2 = a^2$:
$$d^2 - (b - d)^2 = a^2 - c^2$$
$$d^2 - (b^2 - 2bd + d^2) = a^2 - c^2$$
$$d^2 - b^2 + 2bd - d^2 = a^2 - c^2$$
$$-b^2 + 2bd = a^2 - c^2$$
$$2bd = a^2 + b^2 - c^2$$

(B) Note that $4b^2d^2 = (2bd)^2 = (a^2 + b^2 - c^2)^2$.

$$A = \frac{bh}{2} = \sqrt{\frac{b^2h^2}{4}} = \sqrt{\frac{b^2(a^2 - d^2)}{4}} = \sqrt{\frac{a^2b^2 - b^2d^2}{4}} = \sqrt{\frac{4}{4}\left(\frac{a^2b^2 - b^2d^2}{4}\right)}$$

$$= \sqrt{\frac{4a^2b^2 - 4b^2d^2}{16}} = \sqrt{\frac{4a^2b^2 - (a^2 + b^2 - c^2)^2}{16}}$$

(C) Main ideas: Let $x = 2ab$ and $y = a^2 + b^2 - c^2$. Use $x^2 - y^2 = (x + y)(x - y)$.

(D) Main ideas: $(a + b)^2 = a^2 + 2ab + b^2$ and $(a - b)^2 = a^2 - 2ab + b^2$. Note that $-(a - b)^2 = -a^2 + 2ab - b^2$.

(E) Main ideas: Let $t = a + b$ and $u = a - b$. Use $t^2 - c^2 = (t + c)(t - c)$ and $c^2 - u^2 = (c + u)(c - u)$.

(F) Main ideas: $P = a + b + c$, $P - 2a = b + c - a$, $P - 2b = a + c - b$, and $P - 2c = a + b - c$.

$$A = \sqrt{\frac{(a + b + c)(a + b - c)(a + c - b)(b + c - a)}{16}} = \sqrt{\frac{P}{2}\left(\frac{P - 2c}{2}\right)\left(\frac{P - 2b}{2}\right)\left(\frac{P - 2a}{2}\right)}$$

$$A = \sqrt{\frac{P}{2}\left(\frac{P}{2} - \frac{2c}{2}\right)\left(\frac{P}{2} - \frac{2b}{2}\right)\left(\frac{P}{2} - \frac{2a}{2}\right)} = \sqrt{\frac{P}{2}\left(\frac{P}{2} - c\right)\left(\frac{P}{2} - b\right)\left(\frac{P}{2} - a\right)}$$

15. (A) Yes. (B) "If two triangles have the same area, they are congruent." The converse is false. It is possible for two triangles to have the same area, yet have different sides, as the diagram below illustrates. The area of each triangle below is 12, yet these two triangles are clearly not congruent.

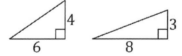

(C) "If two triangles are not congruent, they do not have the same area." The inverse is false for the same reason that the converse is false.

(D) "If two triangles do not have the same area, they are not congruent." The contra-positive is true because two congruent triangles would have the same area.

Chapter 7 Answers

1. (A) $BD = 4\sqrt{3}$. This is equivalent to $\frac{12}{\sqrt{3}}$ since $\frac{12}{\sqrt{3}} = \frac{12}{\sqrt{3}}\frac{\sqrt{3}}{\sqrt{3}} = \frac{12\sqrt{3}}{3} = 4\sqrt{3}$. Check: AB/BD $= \frac{12}{4\sqrt{3}} = \frac{3}{\sqrt{3}} = \frac{3}{\sqrt{3}}\frac{\sqrt{3}}{\sqrt{3}} = \frac{3\sqrt{3}}{3} = \sqrt{3}$ and BC/AB $= \frac{12\sqrt{3}}{12} = \sqrt{3}$. In a 30°-60°-90° triangle, the side opposite to 60° is $\sqrt{3}$ times the side opposite to 30°. Notes: $\angle BAD = \angle CAD = 30°$. See the left diagram below.

(B) $BM = 6\sqrt{3}$. Check: BC/AB $= \frac{12\sqrt{3}}{12} = \sqrt{3}$. In a 30°-60°-90° triangle, the side opposite to 60° is $\sqrt{3}$ times larger than the side opposite to 30°. Notes: BM = CM. See the right diagram below.

(C) No. $\overline{AD}$ is the angle bisector. $\overline{AM}$ is the median. $\overline{AD}$ and $\overline{AM}$ are two different line segments since they intersect $\overline{BC}$ in different places: BM > BD since $6\sqrt{3} > 4\sqrt{3}$.

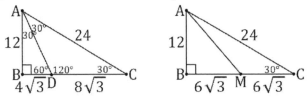

2. (A) $BD = 9$. Check: AB/BD = 18/9 = 2 and BC/AB = 36/18 = 2. In a 30°-60°-90° triangle, the hypotenuse is twice the side opposite to 30°. See the left diagram below. Area check: Since ΔABC is a right triangle, $A = \frac{1}{2}(18)(18\sqrt{3}) = 162\sqrt{3}$, which agrees with $A = \frac{1}{2}(36)(9\sqrt{3}) = 162\sqrt{3}$ using BC = 36 as the base and AD = $9\sqrt{3}$ as the height.

(B) BM = 18. Notes: BM = CM. See the right diagram below.

(C) BM − BD = 18 − 9 = 9.

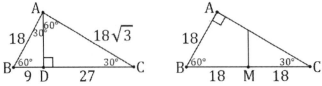

3. AC = 15. Check: BD/AB = 12/20 = 3/5 and DC/AC = 9/15 = 3/5.

4. FG = 7.5. Check: FG/EF = 7.5/18 = 15/36 = 5/12 and GH/EH = 20/48 = 5/12.

5. PS = 12 and QS = 9. Checks: PS + QS = 21, PS/PR = 12/24 = 1/2 and QS/QR = 9/18 = 1/2. Note: One way to solve this problem is to begin with $\frac{QS}{18} = \frac{21-QS}{24}$ and cross multiply to get 24QS = (18)(21) − 18QS.

6. VW = 9/5 = 1.8 and VX = 8/5 = 1.6. Checks: VW/UV = 1.8/3.6 = 1/2 agrees with TW/TU = 3/6 = 1/2, and UX/TU = 2/6 = 1/3 agrees with VX/TV = 1.6/4.8 = 1/3. Notes: One way to solve this problem is to begin with $\frac{VW}{2+VX} = \frac{3}{6}$ and $\frac{2}{6} = \frac{VX}{3+VW}$, and cross multiply to get 6VW = (2)(3) + 3VX and (2)(3) + 2VW = 6VX. Divide both sides of the first equation by 6 and replace VW with that expression in the other equation.

7. DE = 12, EG = 9, and CG = 20. Notes: BC = 2DE, BG = 2EG, and CG = 2DG.

8. GS = 12, GT = 11, and RT = 33. Notes: GR = 2GT, RT = GR + GT = 3GT, GP = 2GS, and PS = PG + GS.

9. AO = BO = 10 and AB = $10\sqrt{3}$. Checks: $AF^2 + FO^2 = 8^2 + 6^2 = 64 + 36 = 100 = 10^2 = AO^2$ and $AD^2 + DO^2 = (5)^2(\sqrt{3})^2 + 5^2 = 25(3) + 25 = 75 + 25 = 100 = 10^2 = AO^2$. Notes: AF = CF = 8 and AD = BD = $5\sqrt{3}$ since D and F are midpoints. AO = BO = CO since O is the circumcenter of $\triangle$ABC.

10. LO = MO = $\sqrt{5}$ and OP = $\sqrt{3}$. Checks: $MN^2 + NO^2 = 1^2 + 2^2 = 1 + 4 = 5 = (\sqrt{5})^2 = MO^2$ and $MP^2 + OP^2 = (\sqrt{2})^2 + (\sqrt{3})^2 = 2 + 3 = 5 = (\sqrt{5})^2 = MO^2$. Notes: LN = MN = 1 and KP = MP = $\sqrt{2}$ since N and P are midpoints. KO = LO = MO since O is the circumcenter of $\triangle$KLM (even though point O lies outside of $\triangle$KLM).

11. AC = 15. Check: A = bh/2 = (AB)(CD)/2 = (10)(12)/2 = 60 agrees with A = bh/2 = (AC)(BE)/2 = (15)(8)/2 = 60. Note: Any side may be treated as the base.

12. HL = 10, HK = 5, KN = 11, KP = 4, LP = 13, MP = 39/4 = 9.75, KM = 55/4 = 13.75, MN = 33/4 = 8.25, LM = 65/4 = 16.25, and KL = $\sqrt{185}$. Similarity checks: HP:KP:HK = 3:4:5, HN:LN:HL = 6:8:10 = 3:4:5, MN:KN:KM = 8.25:11:13.75 = 3:4:5 (divide 8.25, 11, and 13.75 each by 2.75), and MP:LP:LM = 9.75:13:16.25 = 3:4:5 (divide 9.75, 13, and 16.25 each by 3.25). Pythagorean checks: $3^2 + 4^2 = 9 + 16 = 25 = 5^2$, $6^2 + 8^2 = 36 + 64 = 100 = 10^2$, $8.25^2 + 11^2 = 68.0625 + 121 = 189.0625 = 13.75^2$, $9.75^2 + 13^2 = 95.0625 + 169 = 264.0625 = 16.25^2$, and $8^2 + 11^2 = 64 + 121 = 185$.

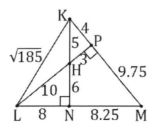

13. α = 24°, δ = 52°, γ = 104°, θ = 24°, φ = 94°, η = 62°, ρ = 28°, σ = 86°, τ = 66°, β = 52°, μ = 100°, ν = 28°, ξ = 62°, ω = 80°, ψ = 38°, κ = 66°, λ = 76°, and χ = 38°. No two triangles are congruent or similar since no two triangles have the same three interior angles. No line segment is an inradius since no line segment is perpendicular to a side of ΔTUV. Notes: α = θ, ρ = ν, and χ = ψ because $\overline{TX}$, $\overline{UY}$, and $\overline{VW}$ are angle bisectors. δ = β, κ = τ, and η = ξ are vertical angles. φ & σ, μ & ω, and γ and λ are supplementary angles. The three interior angles of each triangle (like θ, φ, and η) add up to 180°. Since ∠UTV = 48° and ∠TUV = 56°, it follows that ∠UVT = 76° (since these add up to 180°). **Tip**: Start with ΔUTX and solve for μ.

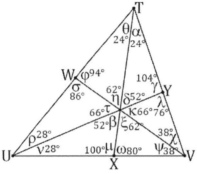

14. ∠1 = ∠6 = 42°, ∠2 = ∠3 = 32°, ∠4 = ∠5 = 16°, ∠7 = ∠12 = 48°, ∠8 = ∠9 = 58°, and ∠10 = ∠11 = 74°. ΔADI ≅ ΔAFI, ΔBDI ≅ ΔBEI, and ΔCEI ≅ ΔCFI via ASA, AAS, or SAS. $\overline{DI}$ ≅ $\overline{EI}$ ≅ $\overline{FI}$ are inradii. Notes: ∠1 = ∠6, ∠2 = ∠3, and ∠4 = ∠5 because $\overline{AI}$, $\overline{BI}$, and $\overline{CI}$ are angle bisectors. Each pair of acute angles in each right triangle (such as ∠1 and ∠7) add up to 90°. ∠7 thru ∠12 add up to 360°. Since ∠ABC = 64° and ∠ACB = 32°, it follows that ∠BAC = 84° (since these add up to 180°).

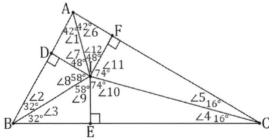

15. α = 34°, δ = λ = 70°, β = γ = 56°, θ = ν = 54°, φ = μ = 36°, and κ = 20°. ΔKNO ≅ ΔLNO, ΔLOP ≅ ΔMOP, and ΔKOQ ≅ ΔMOQ via SAS (one shared side, one 90° angle, and one side that is bisected by a midpoint). $\overline{KO}$ ≅ $\overline{LO}$ ≅ $\overline{MO}$ are circumradii. Notes: Each pair of acute angles in each right triangle (such as θ and φ) add up to 90°. The

angles in the center add up to 360° since they form a full circle. $2\varphi + 2(34°) + 2(20°)$ = 180° because the three interior angles of ΔKLM add up to 180°.

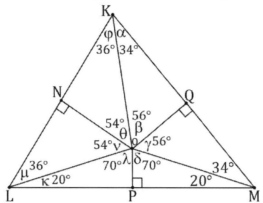

16. $\angle 1 = \angle 8 = 40°$, $\angle 2 = \angle 7 = 50°$, $\angle 3 = \angle 10 = 72°$, $\angle 4 = \angle 9 = 18°$, $\angle 5 = \angle 12 = 32°$, and $\angle 6 = \angle 11 = 58°$. ΔHRU~ΔHTV~ΔSTU~ΔRSV (these are 40°-50°-90°), ΔHSU~ΔHTW~ΔRSW~ΔRTU (these are 18°-72°-90°), and ΔHRW~ΔHSV~ΔSTW~ΔRTV (these are 32°-58°-90°) according to AA. (These sets of triangles are similar, but not congruent.) Notes: Each pair of acute angles in each right triangle (such as $\angle 1$ and $\angle 2$) add up to 90°. The central angles add up to 360°. $\angle 1 + \angle 4 + \angle 5 + \angle 8 + \angle 9 + \angle 12 = 180°$ because the three interior angles of ΔRST add up to 180°. **Tip:** Remember to look at the larger right triangles, too, such as ΔRTU. For example, since $\angle SRT = 72°$, ΔRTU makes it easy to find $\angle 9$. Then knowing $\angle RTS = 58°$ makes it easy to find $\angle 8$.

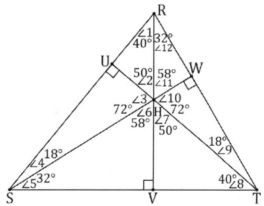

17. (A) $h = \frac{L\sqrt{3}}{2}$. For an equilateral triangle, the angle bisector, perpendicular bisector, median, and altitude are one and the same. The line segment that bisects an interior angle also bisects the opposite side and is perpendicular to the opposite side. Notes: Divide the equilateral triangle in half as shown on the next page. This creates two 30°-

60°-90° triangles. The hypotenuse is L, the side opposite to the 30° angle is L/2, and the side opposite to the 60° angle is $\frac{L\sqrt{3}}{2}$, which is the height of the equilateral triangle. The ratio of the sides of the 30°-60°-90° triangle is $\frac{1}{2}:\frac{\sqrt{3}}{2}:1$. This ratio is equivalent to 1:$\sqrt{3}$:2 (to see this, multiply $\frac{1}{2}$, $\frac{\sqrt{3}}{2}$, and 1 each by 2).

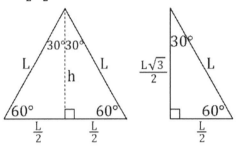

(B) FG $=\frac{L\sqrt{3}}{6}$. The incenter, circumcenter, orthocenter, and centroid are the same for an equilateral triangle. The centroid (G) is one-third of a median (AF, BE, or CD). Since a median has a length of h $=\frac{L\sqrt{3}}{2}$, the centroid lies $\frac{h}{3}=\frac{L\sqrt{3}}{6}$ from sides $\overline{BC}$, $\overline{AC}$, and $\overline{AB}$. Notes: AG = BG = CG $=\frac{L\sqrt{3}}{3}$ such that AG = 2FG, BG = 2GE, and CG = 2DG. Since the centroid is the same as the incenter for an equilateral triangle, the distance from the centroid to one side equals the distance from the incenter to one side.

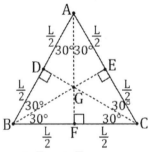

(C) AG $=\frac{L\sqrt{3}}{3}$. Notes: AG = 2FG = $2\frac{L\sqrt{3}}{6}=\frac{L\sqrt{3}}{3}$.

(D) FG $=\frac{L\sqrt{3}}{6}$. This is the same as the answer to Part B.

(E) DE $=\frac{L}{2}$. Note: BC = 2DE = L.

(F) A $=\frac{1}{2}$(b)(h) $=\frac{1}{2}$(L)$\left(\frac{L\sqrt{3}}{2}\right)=\frac{L^2\sqrt{3}}{4}$. Notes: b = BC = L and h = AF $=\frac{L\sqrt{3}}{2}$.

(G) A $=\frac{1}{2}$(P)(r) $=\frac{1}{2}$(3L)$\left(\frac{L\sqrt{3}}{6}\right)=\frac{3L^2\sqrt{3}}{12}=\frac{L^2\sqrt{3}}{4}$. Note: r = FG $=\frac{L\sqrt{3}}{6}$.

(H) $A = \frac{L^2\sqrt{3}}{24}$. Note: The six small triangles below on the left have the same area. Divide the area of the complete triangle by six.

(I) $A = \frac{L^2\sqrt{3}}{16}$. Note: The four small triangles below on the right are congruent. Divide the area of the complete triangle by four.

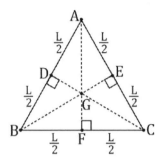 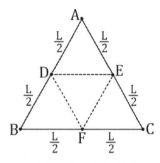

18. (A) The orthocenter lies at point B, where altitudes AB, BC, and BD intersect, as shown below. $AB = BC = 1$ and $BD = \frac{\sqrt{2}}{2}$. Notes: Altitude AB is perpendicular to BC and passes through A, altitude BC is perpendicular to AB and passes through C, and altitude BD is perpendicular to AC and passes through B. Since all three altitudes intersect at point B, this makes point B the orthocenter. $AC = \sqrt{2}$ because the sides of a 45°-45°-90° triangle come in the ratio $1:1:\sqrt{2}$. Since altitude BD is perpendicular to AC, $\angle ABD = \angle CBD = 45°$. This shows that altitude BD is also an angle bisector. Since $AB = BC$, angle bisector BD is also a median: $AD = CD = \frac{\sqrt{2}}{2}$. Since $\triangle BCD$ is a 45°-45°-90° triangle, $BD = CD = \frac{\sqrt{2}}{2}$.

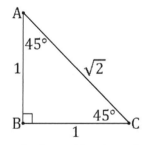 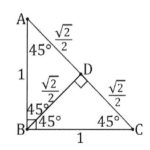

(B) The circumcenter is the midpoint of the hypotenuse (point D shown above) since point D is equidistant from the three vertices: $AD = BD = CD = \frac{\sqrt{2}}{2}$. The circumradius equals $\frac{\sqrt{2}}{2}$.

(C) $BG = \frac{\sqrt{2}}{3}$. Notes: $BG = 2DG$. $DG = \frac{\sqrt{2}}{6}$. $BG + DG = \frac{\sqrt{2}}{3} + \frac{\sqrt{2}}{6} = \frac{2\sqrt{2}}{6} + \frac{\sqrt{2}}{6} = \frac{3\sqrt{2}}{6} = \frac{\sqrt{2}}{2} = BD$.

BG is 2/3 of the distance from B to D, while DG is 1/3 of the distance from D to B.

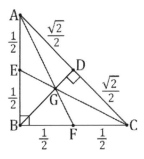

(D) $\alpha = 45°$, $\delta = 67.5°$, $\theta = 22.5°$, and $\varphi = 67.5°$.

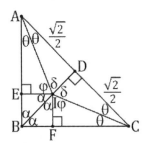

(E) $r = DI = EI = FI = \frac{2-\sqrt{2}}{2}$. The answer is equivalent to $\frac{1}{2+\sqrt{2}}$ because $\frac{1}{2+\sqrt{2}} \frac{2-\sqrt{2}}{2-\sqrt{2}} = \frac{2-\sqrt{2}}{4-2}$ $= \frac{2-\sqrt{2}}{2}$. Notes: Point I is NOT a midpoint of BD (so it would be incorrect to divide BD by 2 to find BI), even though it may appear that way in the diagram above. A simple way to find the inradius is to first find the area of ΔABC: $A = \frac{1}{2}bh = \frac{1}{2}(1)(1) = \frac{1}{2}$. The perimeter is $P = 1 + 1 + \sqrt{2} = 2 + \sqrt{2}$. The area equals one-half the inradius times the perimeter: $A = \frac{1}{2}Pr$ becomes $\frac{1}{2} = \frac{1}{2}(2 + \sqrt{2})r$, which simplifies to $r = \frac{1}{2+\sqrt{2}} = \frac{2-\sqrt{2}}{2}$.

19. (A) The orthocenter lies at point B, where altitudes AB, BC, and BD intersect, as shown on the next page. $AB = 1$, $BC = \sqrt{3}$, and $BD = \frac{\sqrt{3}}{2}$. Notes: Since all three altitudes intersect at point B, this makes point B the orthocenter. $BC = \sqrt{3}$ and $AC = 2$ because the sides of a 30°-60°-90° triangle come in the ratio $1{:}\sqrt{3}{:}2$. $BD = \frac{\sqrt{3}}{2}$, $CD = \frac{3}{2}$, and $BC = \sqrt{3}$ because ΔBCD is a 30°-60°-90° triangle. Multiply $\frac{\sqrt{3}}{2}$, $\frac{3}{2}$, and $\sqrt{3}$ each by $\frac{2}{\sqrt{3}}$ to see that the ratio $\frac{\sqrt{3}}{2}{:}\frac{3}{2}{:}\sqrt{3}$ is equivalent to $1{:}\sqrt{3}{:}2$.

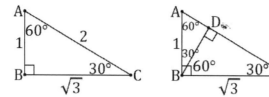

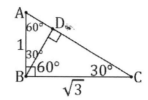

(B) The circumcenter is the midpoint of the hypotenuse (point M shown below) since point M is equidistant from the three vertices: $AM = BM = CM = 1$. The circumradius equals 1. Note: One way to prove that $BM = 1$ is to divide the obtuse triangle below into two congruent 30°-60°-90° triangles.

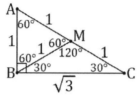

(C) $BG = \frac{2}{3}$. Notes: $BG = 2GM$. $GM = \frac{1}{3}$. $BG + GM = \frac{2}{3} + \frac{1}{3} = \frac{3}{3} = 1 = BM$. BG is 2/3 of the distance from B to M, while GM is 1/3 of the distance from B to M.

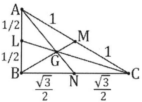

(D) $\alpha = 15°, \beta = 75°, \delta = 45°, \theta = 30°, \varphi = 60°,$ and $\gamma = 75°$.

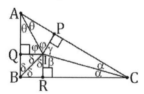

(E) $r = IP = IQ = IR = \frac{\sqrt{3}-1}{2}$. The answer is equivalent to $\frac{\sqrt{3}}{3+\sqrt{3}}$ and to $\frac{1}{\sqrt{3}+1}$ because

$\frac{\sqrt{3}}{3+\sqrt{3}}\frac{3-\sqrt{3}}{3-\sqrt{3}} = \frac{3\sqrt{3}-3}{9-3} = \frac{3(\sqrt{3}-1)}{6} = \frac{\sqrt{3}-1}{2}$ and since $\frac{1}{\sqrt{3}+1}\frac{\sqrt{3}}{\sqrt{3}} = \frac{\sqrt{3}}{3+\sqrt{3}}$. Notes: Points B, I, and P are NOT collinear, and I is NOT equidistant from B and P. A simple way to find the inradius is to first find the area of $\triangle ABC$: $A = \frac{1}{2}bh = \frac{1}{2}(\sqrt{3})(1) = \frac{\sqrt{3}}{2}$. The perimeter is

$P = 1 + \sqrt{3} + 2 = 3 + \sqrt{3}$. The area equals $A = \frac{1}{2}Pr$, such that $\frac{\sqrt{3}}{2} = \frac{1}{2}(3+\sqrt{3})r$, which

simplifies to $r = \frac{\sqrt{3}}{3+\sqrt{3}} = \frac{\sqrt{3}}{3+\sqrt{3}}\frac{3-\sqrt{3}}{3-\sqrt{3}} = \frac{3\sqrt{3}-3}{9-3} = \frac{3(\sqrt{3}-1)}{6} = \frac{\sqrt{3}-1}{2}$.

20. (A) FH = 9/4 = 2.25. Notes: AF = 4 since $BF^2 + AF^2 = 3^2 + 4^2 = 9 + 16 = 25 = 5^2 = AB^2$. ∠CAF ≅ ∠CBE because ∠CAF & ∠AHE are complements, ∠AHE and ∠BHF are vertical angles, and ∠BHF and ∠CBE are complements. ΔBFH~ΔACF according to AA. FH/BF = CF/AF becomes $\frac{FH}{3} = \frac{3}{4}$. Cross multiply: 4FH = 9 leads to FH = 9/4.

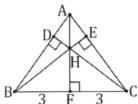

(B) AO = BO = CO = 25/8 = 3.125. Notes: Apply the Pythagorean theorem to ΔBLO: $3^2 + LO^2 = BO^2$. Since AO = BO = CO, this becomes $3^2 + LO^2 = AO^2$. The height is h = 4 = AO + LO, such that LO = 4 − AO. Replace LO with 4 − AO in $3^2 + LO^2 = AO^2$ to get $3^2 + (4 - AO)^2 = AO^2$. Apply algebra: $9 + 16 - 8AO + AO^2 = AO^2$. This simplifies to 25 = 8AO, which gives 25/8 = AO.

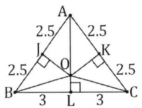

(C) GL = 4/3. Notes: AG = 2GL= 8/3 and AG + GL = h = 8/3 + 4/3 = 4.

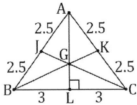

(D) r = IM = IN = IP = 3/2 = 1.5. Notes: Apply the triangle bisector theorem to ΔABP: $\frac{3}{IP} = \frac{5}{AI}$. Cross multiply: 3AI = 5IP. The height is AI + IP = 4, such that AI = 4 − IP. Replace AI with 4 − IP: 3(4 − IP) = 5IP. Distribute: 12 − 3IP = 5IP. This simplifies to 12 = 8IP, which gives 12/8 = 3/2 = 1.5 = IP. An alternative method is to use $A = \frac{1}{2}Pr$.

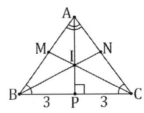

21. (A) HM $= 3/2 = 1.5$. Altitude $\overline{BH}$ is perpendicular to side $\overline{AC}$; side $\overline{AC}$ was extended with $\overline{AY}$ in order to find the point on line $\overleftrightarrow{AC}$ for which a perpendicular line intersects vertex B. Similarly, altitude $\overline{CH}$ is perpendicular to side $\overline{AB}$ (which was extended to Z) and intersects vertex C. Altitude $\overline{HM}$ is perpendicular to side $\overline{BC}$ and intersects vertex A. The three altitudes ($\overline{BH}$, $\overline{CH}$, and $\overline{HM}$) intersect at the orthocenter (H). To find any angles or sides outside of ΔABC, a good place to start is to note that $\angle CAZ = 60°$ since $\angle BAM + \angle CAM + \angle CAZ = 180°$. It follows that $\Delta AMC \cong \Delta ACZ$ (by ASA) and similarly $\Delta ACZ \cong \Delta AHZ$ (by ASA). Therefore, AH $= 1$ and HM $=$ AH $+$ AM $= 1 + 1/2 = 3/2 = 1.5$. ΔBCH is an equilateral triangle. Point A is the centroid of ΔBCH (but NOT the centroid of ΔABC for Part C), located $1/3$ of the height (HM) of ΔBCH.

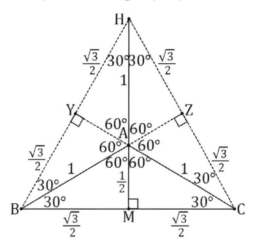

(B) AO $=$ BO $=$ CO $= 1$. Notes: ΔAOV is a 30°-60°-90° triangle. The side opposite to the 30° angle is $1/2$. The hypotenuse (AO) is therefore 1. $\Delta AOV \cong \Delta BOV$ and $\Delta AOW \cong \Delta COW$ according to SAS.

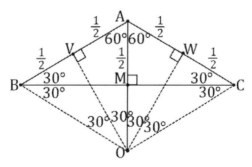

(C) GM $= 1/6$. Notes: AG $=$ 2GM $= 1/3$. AG $+$ GM $= 1/3 + 1/6 = 2/6 + 1/6 = 3/6 = 1/2$. The medians are drawn on the following page.

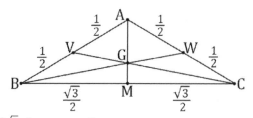

(D) $r = IM = IT = IU = \frac{2\sqrt{3}-3}{2} = \sqrt{3} - \frac{3}{2}$. The answer is equivalent to $\frac{3}{4\sqrt{3}+6}$, $\frac{3}{2(2\sqrt{3}+3)}$, and

$\frac{\sqrt{3}}{2(2+\sqrt{3})}$ because $\frac{3}{4\sqrt{3}+6} \frac{4\sqrt{3}-6}{4\sqrt{3}-6} = \frac{12\sqrt{3}-18}{48-36} = \frac{6(2\sqrt{3}-3)}{12} = \frac{2\sqrt{3}-3}{2}$, $\frac{3}{2(2\sqrt{3}+3)} = \frac{3}{4\sqrt{3}+6}$, and $\frac{\sqrt{3}}{2(2+\sqrt{3})}$

$= \frac{\sqrt{3}}{2(2+\sqrt{3})} \frac{\sqrt{3}}{\sqrt{3}} = \frac{3}{2(2\sqrt{3}+3)}$. The preferred forms of the answer are $\frac{2\sqrt{3}-3}{2}$ or $\sqrt{3} - \frac{3}{2}$ because

these have rational denominators. Notes: The perimeter is $P = 1 + 1 + \sqrt{3} = 2 + \sqrt{3}$.

The area is $A = \frac{1}{2}bh = \frac{1}{2}(\sqrt{3})(\frac{1}{2}) = \frac{\sqrt{3}}{4}$. An alternative formula for the area is $A = \frac{1}{2}Pr$,

such that $\frac{\sqrt{3}}{4} = \frac{1}{2}(2 + \sqrt{3})r$. Multiply both sides by 2 to get $\frac{\sqrt{3}}{2} = (2 + \sqrt{3})r$. Divide by

$(2+\sqrt{3})$ on both sides to get $r = \frac{\sqrt{3}}{2(2 + \sqrt{3})} = \frac{\sqrt{3}}{2(2+\sqrt{3})} \frac{\sqrt{3}}{\sqrt{3}} = \frac{3}{2(2\sqrt{3}+3)} = \frac{3}{4\sqrt{3}+6} = \frac{3}{4\sqrt{3}+6} \frac{4\sqrt{3}-6}{4\sqrt{3}-6}$

$= \frac{12\sqrt{3}-18}{48-36} = \frac{6(2\sqrt{3}-3)}{12} = \frac{2\sqrt{3}-3}{2}$. Alternative method: Apply the triangle bisector theorem

to $\triangle ABM$: $\frac{\sqrt{3}}{2IM} = \frac{1}{AI}$. Cross multiply: $AI\sqrt{3} = 2IM$. The height is $AI + IM = \frac{1}{2}$, such that AI

$= \frac{1}{2} - IM$. Replace AI with $\frac{1}{2} - IM$: $(\frac{1}{2} - IM)\sqrt{3} = 2IM$. Distribute: $\frac{\sqrt{3}}{2} - IM\sqrt{3} = 2IM$.

This simplifies to $\frac{\sqrt{3}}{2} = IM(2 + \sqrt{3})$, which gives $IM = \frac{\sqrt{3}}{2(2+\sqrt{3})}$.

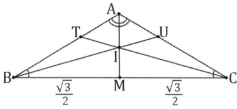

22. For the acute triangle, the incenter, circumcenter, centroid, and orthocenter all lie inside of the triangle, as shown below.

For the right triangle, the incenter and centroid lie inside of the triangle, the circumcenter lies on the triangle at the midpoint of the hypotenuse, and the orthocenter lies on the triangle at the vertex opposite to the hypotenuse, as shown on the next page.

For the obtuse triangle, the incenter and centroid lie inside of the triangle, and the circumcenter and orthocenter lie outside of the triangle, as shown below.

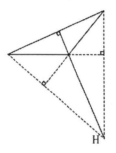

23. The area of ΔABC is 72, the area of ΔADG is 12, the area of ΔACE is 36, the area of ΔDEF is 18, the area of ΔABG is 24, and the area of ΔDEG is 6. Notes: ΔADF ≅ ΔBEF ≅ ΔDEF ≅ ΔCDE. The area of ΔCDE and ΔDEF are each 18, and the area of ΔABC is 4×18 = 72. ΔAFG, ΔBFG, ΔBEG, ΔCEG, ΔCDG, and ΔADG have the same area. The area of ΔADG is 72÷6 = 12. The area of ΔACE is 3×12 = 36 because ΔCEG, ΔCDG, and ΔADG have the same area. The area of ΔABG is 2×12 = 24 because ΔAFG and ΔBFG have the same area. Point G is the centroid of both ΔABC and ΔDEF (since the medians of ΔABC are the medians of ΔDEF; use the property of similarity established in Problems 30-32). Thus, point G divides ΔDEF into 6 smaller congruent triangles, each with area 18÷6 = 3. The area of ΔDEG is 2×3 = 6.

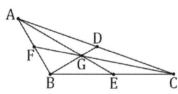 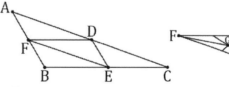

24. $A = 6\sqrt{6}$. The three altitudes are $\frac{12\sqrt{6}}{7}$, $2\sqrt{6}$, and $\frac{12\sqrt{6}}{5}$. The inradius is $r = \frac{2\sqrt{6}}{3}$. Notes: The perimeter is $P = 5 + 6 + 7 = 18$, such that $P/2 = 9$. Use Heron's formula (Chapter 6): $A = \sqrt{9(9-7)(9-6)(9-5)} = \sqrt{9(2)(3)(4)} = \sqrt{9}\sqrt{6}\sqrt{4} = 3(2)\sqrt{6} = 6\sqrt{6}$. For each altitude, area equals one-half base times the height: $\frac{1}{2}(7)\left(\frac{12\sqrt{6}}{7}\right) = 6\sqrt{6}, \frac{1}{2}(6)(2\sqrt{6}) = 6\sqrt{6}$, and $\frac{1}{2}(5)\left(\frac{12\sqrt{6}}{5}\right) = 6\sqrt{6}$. Area also equals one-half the perimeter times inradius: $6\sqrt{6} = 9r$, such that $\frac{6\sqrt{6}}{9} = \frac{2\sqrt{6}}{3} = r$.

25. Main ideas: ∠EFK ≅ ∠EJK (both are right), ∠GEH ≅ ∠HEI (since $\overline{EH}$ bisects ∠GEI), and side $\overline{EK}$ is shared. ΔEFK ≅ ΔEJK according to AAS. $\overline{FK}$ ≅ $\overline{JK}$ by the CPCTC.

26. Main ideas: ∠BDE ≅ ∠ADC (vertical angles) and ∠DBE ≅ ∠ACD (alternate interior angles). ΔBDE~ΔACD according to AA. The sides of similar triangles come in the same proportion: DE/BE = AD/AC. Since ∠DBE ≅ ∠ACD (alternate interior angles) and ∠DBE ≅ ∠ABD ($\overline{BC}$ bisects ∠ABE), ∠ACD ≅ ∠ABD, such that ΔABC is isosceles. It follows that $\overline{AB}$ ≅ $\overline{AC}$. Replace AC with AB in the previous proportion: DE/BE = AD/AB. The proof remains valid even if $\overline{AC}$ and $\overline{CD}$ are erased from the diagram; $\overline{AC}$ and $\overline{CD}$ exist whether or not they are drawn.

27. Main ideas: $\overline{AB}$ is the line segment that is perpendicular to $\overline{BC}$ and passes through vertex A, so $\overline{AB}$ is one altitude. $\overline{BC}$ is the line segment that is perpendicular to $\overline{AB}$ and passes through vertex C, so $\overline{BC}$ is another altitude. $\overline{BD}$ is perpendicular to $\overline{AC}$ and passes through vertex B, so $\overline{BD}$ is another altitude. All three altitudes ($\overline{AB}$, $\overline{BC}$, and $\overline{BD}$) intersect at point B. Therefore, vertex B is the orthocenter.

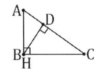

Points E, F, and O are the midpoints of $\overline{AB}$, $\overline{BC}$, and $\overline{AC}$. EO and FO are the perpendicular bisectors. (The third perpendicular bisector, the one perpendicular to $\overline{AC}$, has zero length; it extends from point O to point O.) In order to show that point O (which is the midpoint of the hypotenuse) is the circumcenter, it is necessary to show that $\overline{AO}$, $\overline{CO}$, and $\overline{BO}$ are all congruent. $\overline{AO}$ ≅ $\overline{CO}$ because point O is the midpoint of $\overline{AC}$. In order to show that $\overline{BO}$ ≅ $\overline{AO}$, first show that ΔAEO ≅ ΔBEO and then use the CPCTC. ΔAEO ≅ ΔBEO because $\overline{AE}$ ≅ $\overline{BE}$ (since E is the midpoint of $\overline{AB}$), ∠AEO ≅ ∠BEO (both are right according to the midsegment theorem), and side $\overline{EO}$ is shared (SAS).

28. Main ideas: Let I be the point where $\overline{AI}$ and $\overline{BI}$ intersect. Let D and F be the points where line segments from point I are perpendicular to sides $\overline{AB}$ and $\overline{AC}$. ∠ADI ≅ ∠AFI (both are right), ∠DAI ≅ ∠FAI (since $\overline{AI}$ bisects ∠DAF), and $\overline{AI}$ is shared by ΔADI and ΔFAI. Therefore, ΔADI ≅ ΔAFI according to AAS and $\overline{DI}$ ≅ $\overline{FI}$ according to the CPCTC. Similar arguments show that ΔBDI ≅ ΔBEI and ΔCEI ≅ ΔCFI, and that $\overline{DI}$ ≅ $\overline{EI}$ and $\overline{EI}$ ≅ $\overline{FI}$. This shows that $\overline{DI}$ ≅ $\overline{EI}$ ≅ $\overline{FI}$. The circumradius is r = DI = EI = FI. Point I, the circumcenter, is equidistant from points D, E, and I.

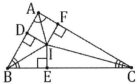

Since point I is equidistant from $\overline{AB}$ and $\overline{AC}$ (since $\overline{DI} \cong \overline{FI}$), I lies on the angle bisector of $\angle DAF$ according to the result of Problem 25. Similarly, I lies on the angle bisectors of $\angle DBE$ and $\angle EFC$ because point I is equidistant from $\overline{AB}$ and $\overline{BC}$ (since $\overline{DI} \cong \overline{EI}$) and point I is equidistant from $\overline{BC}$ and $\overline{AC}$ (since $\overline{EI} \cong \overline{FI}$). Therefore, point I lies on all of the angle bisectors (for $\angle DAF$, $\angle DBE$, and $\angle EFC$).

29. Main ideas: Add up the areas of the six right triangles in the previous diagram.

$$A = \frac{1}{2}(AD)(DI) + \frac{1}{2}(BD)(DI) + \frac{1}{2}(BE)(EI) + \frac{1}{2}(CE)(EI) + \frac{1}{2}(CF)(FI) + \frac{1}{2}(AF)(FI)$$

As discussed in the solution to Problem 28, $r = DI = EI = FI$. Replace DI, EI, and FI with r in the previous equation.

$$A = \frac{1}{2}(AD)r + \frac{1}{2}(BD)r + \frac{1}{2}(BE)r + \frac{1}{2}(CE)r + \frac{1}{2}(CF)r + \frac{1}{2}(AF)r$$

Factor out the 1/2 and the r:

$$A = \frac{1}{2}(AD + BD + BE + CE + CF + AF)r$$

Note that $AD + BD + BE + CE + CF + AF$ is the perimeter, such that $A = \frac{1}{2}Pr$.

30. Main ideas: $\overline{AF} \cong \overline{CF}$ (since F is a midpoint of $\overline{AC}$), $\angle AFD \cong \angle CFP$ (vertical angles), and $\overline{DF} \cong \overline{FP}$ (given). Therefore, $\triangle ADF \cong \triangle CFP$ according to SAS. $\angle DAF \cong \angle FCP$ via the CPCTC. $AF = CF = AC/2$, $\angle DAF \cong \angle FCP$, and $AD = CP = AB/2$ show that $\triangle ADF \sim \triangle ABC$ and $\triangle CFP \sim \triangle ABC$ according to SAS similarity. Each side of $\triangle ABC$ is twice as long as the corresponding side of $\triangle ADF$, such that $BC = 2DF$. $\angle ADF \cong \angle DBE$ because these angles correspond in $\triangle ADF$ and $\triangle ABC$. Since $\angle BDF$ is a supplement to $\angle ADF$ and since $\angle ADF \cong \angle DBE$, it follows that $\angle BDF + \angle DBE = 180°$. $\overline{DF} \parallel \overline{BC}$ according to the parallel postulate (Chapter 1). Note: $\angle ADF$ and $\angle DBE$ are corresponding angles. Alternative proofs: One alternative way to prove the midsegment theorem (or midline theorem) is to prove that BCPD is a parallelogram and then use the properties of a parallelogram. Another way to prove the midsegment theorem is to first prove that the six triangles formed by the medians have equal area (without using the midsegment theorem in the proof; otherwise the subsequent proof of the midsegment theorem would involve circular reasoning) and then use that to prove the midsegment theorem.

31. Main ideas: Use the results of Problem 30: BC = 2DF and $\overline{DF} \parallel \overline{BC}$. BE = CE = BC/2 (since E is the midpoint of $\overline{BC}$). Since DF = BC/2 and CE = BC/2, it follows that CE = DF (from the transitive property). ∠ADF ≅ ∠FEC and ∠AFD ≅ ∠FCE because $\overline{DF} \parallel \overline{BC}$ (make pairs of corresponding angles and apply the transitive property to show this). ΔADF ≅ ΔCEF according to ASA.

32. Main ideas: Use the results of Problem 30: BC = 2DF and $\overline{DF} \parallel \overline{BC}$. ∠FDG ≅ ∠BCG and ∠DFG ≅ ∠CBG are two pairs of alternate interior angles (since $\overline{DF} \parallel \overline{BC}$). ΔDFG~ ΔBCG according to AA. Each side of ΔBCG is twice as long as the corresponding side of ΔDFG, such that BG = 2FG, and CG = 2DG. These results can be used to show that the three medians are concurrent as follows. Consider the two triangles below. Just as BG = 2FG and CG = 2DG were shown for the left diagram, BG = 2FG and AG = 2GE can be shown for the right diagram. The point G where medians $\overline{BF}$ and $\overline{CD}$ intersect is two-thirds of the length from B to F in the left diagram below. The point G where medians $\overline{BF}$ and $\overline{AE}$ intersect is also two-thirds of the length from B to F in the right diagram below. All three medians are therefore concurrent, intersecting at the same point G (which is the centroid).

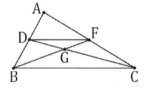 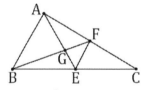

Note: An alternative way to prove that the three medians are concurrent is to set up a coordinate system with the origin at vertex B, write equations for the line of each side of the triangle and each median (using the fact that points D, E, and F are midpoints), and show that the equations for the medians have one common point of intersection.

33. Main ideas: $\overline{AF} \cong \overline{CF}$ since $\overline{BF}$ is a median. ΔAFG and ΔCFG have the same area since they have the same base (AF = CF) and the same height (since point G is the furthest part of each triangle from the base; the height of each triangle is the shortest distance from G to the $\overline{AC}$). (Note that FG is generally NOT the height of either triangle; the height is not marked in the diagram.) Similarly, ΔBEG and ΔCEG have the same area (since $\overline{BE} \cong \overline{CE}$ and they have the same height), and ΔADG and ΔBDG have the same area (since $\overline{AD} \cong \overline{BD}$ and they have the same height). Using similar reasoning, ΔABE and ΔACE have the same area. Combined with previous results, it follows that ΔADG and

ΔBDG have the same area as ΔAFG and ΔCFG. Similarly, ΔABF and ΔCBF have the same area, from which it follows that ΔADG and ΔBDG have the same area as ΔBEG and ΔCEG. Similarly, ΔACD and ΔBCD have the same area, from which it follows that ΔAFG and ΔCFG have the same area as ΔBEG and ΔCEG. Combined together, these results show that all six triangles formed by the three medians have the same area. It follows that ΔACG, ΔBCG, and ΔABG have the same area, since each of these triangles consists of two of the smaller triangles which have equal area.

34. Main ideas: ∠BAP ≅ ∠ABC and ∠ABP ≅ ∠BAC are two pairs of alternate interior angles (since $\overline{PQ} \parallel \overline{BC}$ and $\overline{PR} \parallel \overline{AC}$). ΔABP ≅ ΔABC according to ASA (since side $\overline{AB}$ is shared by the two triangles). Similar arguments show that ΔACQ ≅ ΔABC and ΔBCR ≅ ΔABC. From the CPCTC, $\overline{BP} \cong \overline{BR}$, $\overline{AP} \cong \overline{AQ}$, and $\overline{CR} \cong \overline{CQ}$. This shows that points A, B, and C are midpoints of ΔPQR. Therefore, the altitudes of ΔABC are the perpendicular bisectors of ΔPQR. Since the perpendicular bisectors of ΔPQR are concurrent (as shown in Example 10), the altitudes of ΔABC are concurrent.

35. (A) Yes. The idea is similar to Example 6 in reverse.

(B) "If at least two sides of the triangle are congruent, at least one median is also an angle bisector." The converse is true if it is worded like this. Note that only the angle bisector formed by the congruent sides is guaranteed to be a median; the other two angle bisectors are not medians (unless the triangle is equilateral).

(C) "If none of the medians is also an angle bisector, the triangle is not isosceles." The inverse is true if it is worded like this. See the note for Part B.

(D) "If the triangle is scalene, no median is an angle bisector." The contrapositive is true.

Chapter 8 Answers

1. (A) $2 < c < 8$. (B) $0 < c < 12$. (C) $7 < c < 11$. (D) $9 < c < 33$.

2. $c < a < b$. Note: The missing angle is $180° - 57° - 64° = 59°$. Side c is opposite to $57°$, a is opposite to $59°$, and b is opposite to $64°$.

3. (A) acute (B) obtuse (C) right (D) impossible. Notes: In (A), $9^2 = 81$ and $6^2 + 7^2 = 36 + 49 = 85$ such that $9^2 < 6^2 + 7^2$. In (B), $14^2 = 196$ and $6^2 + 12^2 = 36 + 144 = 180$ such that $14^2 > 6^2 + 12^2$. In (C), $17^2 = 289$ and $8^2 + 15^2 = 64 + 225 = 289$ such that $17^2 = 8^2 + 15^2$. In (D), $9 + 12 = 21 < 24$, which violates the triangle inequality.

4. Main ideas: Since $\overline{BD} \perp \overline{AD}$, line segment $\overline{BD}$ is the shortest possible connection from B to $\overline{AD}$ (as discussed in Example 9 of Chapter 7). Therefore, $AB > BD$. Similarly, since $\overline{CD} \perp \overline{AD}$, line segment $\overline{CD}$ is the shortest possible connection from C to $\overline{AD}$. Therefore, $AC > CD$. Add the two inequalities together: $AB + AC > BD + CD$. Since $BD + CD = BC$, it follows that $AB + AC > BC$.

5. Main ideas: AC is the longest side. $\angle ABC$ is the largest interior angle. On the following page, $\triangle ABC$ in the middle is a right triangle: $\angle ABC = 90°$. According to the Pythagorean theorem, $AC^2 = AB^2 + BC^2$ if $\triangle ABC$ is a right triangle. On the following page, $\triangle ABC$ on the left is an acute triangle. Altitude $\overline{AD}$ divides $\triangle ABC$ into two right triangles. $AC^2 = AD^2 + CD^2$ for $\triangle ACD$. Since $AD < AB$ (see the solution to Problem 4) and $CD < BC$, it follows that $AD^2 + CD^2 < AB^2 + BC^2$. This means that $AC^2 < AB^2 + BC^2$ if $\triangle ABC$ is an acute triangle. On the following page, $\triangle ABC$ on the right is an obtuse triangle. Altitude $\overline{AE}$ makes right triangles $\triangle ABE$ and $\triangle ACE$. $AC^2 = AE^2 + CE^2$ for $\triangle ACE$ and $AB^2 = AE^2 + BE^2$ for $\triangle ABE$. Since $BE + BC = CE$, the first equation may be written as $AC^2 = AE^2 + (BE + BC)^2$. Recall from algebra that $(a + b)^2 = a^2 + 2ab + b^2$.

$$AC^2 = AE^2 + (BE + BC)^2$$
$$AC^2 = AE^2 + BE^2 + 2(BE)(BC) + BC^2$$

Since $AB^2 = AE^2 + BE^2$, this equation becomes:

$$AC^2 = AB^2 + 2(BE)(BC) + BC^2$$

This means that $AC^2 > AB^2 + BC^2$ if $\triangle ABC$ is an obtuse triangle. Note: Since $\angle ABC$ is the largest interior angle, $\angle BAC$ and $\angle ACB$ are definitely acute. The only angle that can be right or obtuse is $\angle ABC$.

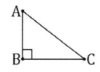 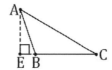

6. Main ideas: The proof for Problem 4 can be adjusted to show that $AB + AC > BC$, $AB + BC > AC$, and $AC + BC > AB$ for any triangle, regardless of which side is longest or shortest. This is the triangle inequality. Subtract AC from both sides of $AB + AC > BC$ to get $AB > BC - AC$. Similarly, subtract BC from both sides of $AB + BC > AC$ to get $AB > AC - BC$. Combine $AB > BC - AC$ and $AB > AC - BC$ together to get $AB > |BC - AC|$.

7. (A) Yes. (B) "If $\angle ABC$ is an obtuse angle, $AC^2 > AB^2 + BC^2$." The converse is true.

(C) "If $AC^2 \leq AB^2 + BC^2$, $\angle ABC$ is not an obtuse angle." The inverse is true.

(D) "If $\angle ABC$ is not an obtuse angle, $AC^2 \leq AB^2 + BC^2$." The contrapositive is true.

Chapter 9 Answers

1. $\angle 1 = 126°$. Check: $126° + 97° + 78° + 59° = 360°$.

2. $\angle DAE = 120°$, $\angle ABC = 60°$, $\angle BCD = 120°$, $AD = 10$, $AB = 20$, $BC = 10$, $h = 5\sqrt{3}$, P $= 60$, and $A = 100\sqrt{3}$. Notes: ΔADF is a 30°-60°-90° triangle. The hypotenuse (AD) is twice the side (DF) that is opposite to the 30° angle. Opposite sides of a parallelogram are congruent. Adjacent interior angles of a parallelogram are supplementary. A = bh.

3. $m = 16$, $P = 64$, and $A = 240$. Notes: $m = \frac{12+20}{2} = \frac{32}{2} = 16$. Draw $\overline{GK}$ to divide the trapezoid into a right triangle and a rectangle. $JK + IK = 8 + 12 = 20 = IJ$. Apply the Pythagorean theorem: $JK^2 + GK^2 = 8^2 + 15^2 = 64 + 225 = 289 = 17^2 = GJ^2$. $A = mh$.

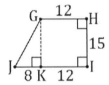

4. $h = 12$, $P = 72$, and $A = 276$. Notes: Check the height with the Pythagorean theorem: $5^2 + 12^2 = 25 + 144 = 169 = 13^2$. $P = 2MN + 2LM$. $A = bh = (MN)(LP)$.

5. $\angle BAD = 120°$, $h = 15\sqrt{3}$, $m = 40$, $d_1 = d_2 = 5\sqrt{91}$, $P = 140$, and $A = 600\sqrt{3}$. Notes: Draw $\overline{AE}$ and $\overline{BF}$ to divide the trapezoid into two right triangles and a rectangle. $CD = DE + EF + CF = 15 + 25 + 15 = 55$. ΔADE is a 30°-60°-90° triangle. $DE:AE:AD = 15:15\sqrt{3}:30 = 1:\sqrt{3}:2$. $m = \frac{25+55}{2} = \frac{80}{2} = 40$. The area is mh. The interior angles add up to $360°$. $d_1 = d_2 = \sqrt{40^2 + h^2} = \sqrt{2275} = \sqrt{(25)(91)} = 5\sqrt{91}$. $h^2 = 225(3) = 675$.

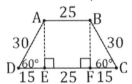

6. $P = 40$ and $A = 96$. Notes: Since the diagonals of a rhombus are perpendicular, they divide the rhombus into four congruent right triangles. $EK + GK = 6 + 6 = 12 = EG$. Check that $HK = 8$ with the Pythagorean theorem: $6^2 + 8^2 = 36 + 64 = 100 = 10^2$. Use $A = \frac{1}{2}d_1 d_2$ with $d_1 = EG = 12$ and $d_2 = FH = 2HK = 2(8) = 16$.

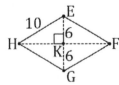

7. $P = 8 + 2\sqrt{3} = 2\left(4 + \sqrt{3}\right)$ and $A = 4 + \frac{5\sqrt{3}}{3}$. Notes: Draw horizontal line segment $\overline{IM}$ and vertical line segments $\overline{IO}$ and $\overline{JN}$ to divide the quadrilateral into one square, two small 30°-60°-90° triangles, and one larger 30°-60°-90° triangle. $KO{:}IO{:}IK = JM{:}IM{:}IJ = 1{:}\sqrt{3}{:}2$. $LN{:}JN{:}JL$ has the same proportions, but $JN = JM + MN = 1 + \sqrt{3}$ such that $LN{:}JN{:}JL = \frac{1+\sqrt{3}}{\sqrt{3}}{:}1 + \sqrt{3}{:}\frac{2+2\sqrt{3}}{\sqrt{3}}$. $\angle KIJ = 30° + 90° + 30° = 150°$. The perimeter is:

$$P = IK + IJ + JL + NL + NO + KO = 2 + 2 + \frac{2 + 2\sqrt{3}}{\sqrt{3}} + \frac{1 + \sqrt{3}}{\sqrt{3}} + \sqrt{3} + 1$$

$$= 5 + \frac{3 + 3\sqrt{3}}{\sqrt{3}} + \sqrt{3} = 5 + \frac{\sqrt{3}(\sqrt{3} + 3)}{\sqrt{3}} + \sqrt{3}$$

$$= 5 + \sqrt{3} + 3 + \sqrt{3} = 8 + 2\sqrt{3} = 2\left(4 + \sqrt{3}\right)$$

The area of ΔKOI is $\frac{1}{2}(1)\sqrt{3} = \frac{\sqrt{3}}{2}$, the area of ΔIJM is $\frac{1}{2}\left(\sqrt{3}\right)(1) = \frac{\sqrt{3}}{2}$, the area of $IMNO$ is $\sqrt{3}\sqrt{3} = 3$, and area of ΔJLN is:

$$\frac{1}{2}\left(\frac{1 + \sqrt{3}}{\sqrt{3}}\right)\left(1 + \sqrt{3}\right) = \frac{1}{2\sqrt{3}}\left(1 + 2\sqrt{3} + 3\right) = \frac{4 + 2\sqrt{3}}{2\sqrt{3}}$$

$$= \frac{2 + \sqrt{3}}{\sqrt{3}} = \frac{2 + \sqrt{3}}{\sqrt{3}}\frac{\sqrt{3}}{\sqrt{3}} = \frac{2\sqrt{3} + 3}{3}$$

The total area is:

$$A = \frac{\sqrt{3}}{2} + \frac{\sqrt{3}}{2} + 3 + \frac{2\sqrt{3} + 3}{3} = \sqrt{3} + 3 + \frac{2\sqrt{3}}{3} + 1$$

$$= 4 + \sqrt{3} + \frac{2\sqrt{3}}{3} = 4 + \frac{3\sqrt{3}}{3} + \frac{2\sqrt{3}}{3} = 4 + \frac{5\sqrt{3}}{3} \text{ or } \frac{12}{3} + \frac{5\sqrt{3}}{3} = \frac{12 + 5\sqrt{3}}{3}$$

8. $P = 66$ and $A = 252$. Notes: $PM = MN = 13$ and $NO = OP = 20$. Check the sides using the Pythagorean theorem:

$$5^2 + 12^2 = 25 + 144 = 169 = 13^2$$

$$12^2 + 16^2 = 144 + 256 = 400 = 20^2$$

$P = 2(13) + 2(20) = 26 + 40 = 66. d_1 = 12 + 12 = 24, d_2 = 5 + 16 = 21, A = \frac{1}{2} d_1 d_2.$

9. $A = 324$. Notes: $L = 18, P = 4L = 4(18) = 72$, and $A = L^2 = 18^2 = 324$.

10. $L = 16, W = 12$, and $d_1 = d_2 = 20$. Notes: $P = 2L + 2W = 2(16) + 2(12) = 32 + 24 = 56$ and $A = LW = (16)(12) = 192$. Use the Pythagorean theorem to find the diagonals: $12^2 + 16^2 = 144 + 256 = 400 = 20^2$. It may help to know the quadratic formula.

11. Main ideas: $\angle 1 \cong \angle 2 \cong \angle 3 \cong \angle 4$ such that $\angle 1 + \angle 2 + \angle 3 + \angle 4 = 4\angle 1 = 360°$ and $\angle 1 = \frac{360°}{4} = 90°$.

12. Main ideas: $\angle BAC + \angle ABC + \angle ACB = 180°$ for $\triangle ABC$ and $\angle CAD + \angle ADC + \angle ACD = 180°$ for $\triangle ACD$. Add these equations together.

$$\angle BAC + \angle ABC + \angle ACB + \angle CAD + \angle ADC + \angle ACD = 360°$$

Plug $\angle BAC + \angle CAD = \angle BAD$ and $\angle ACB + \angle ACD = \angle BCD$ into the previous equation.

$$\angle BAD + \angle ABC + \angle BCD + \angle ADC = 360°$$

13. Main ideas: $\overline{EK} \cong \overline{GK}$ and $\overline{FK} \cong \overline{HK}$ (definition of bisect). $\angle EKH \cong \angle FKG$ (vertical angles). $\triangle EHK \cong \triangle FGK$ according to SAS. $\angle HEK \cong \angle FGK$ according to the CPCTC. $\angle HEK$ and $\angle FGK$ are alternate interior angles, showing that $\overline{EH} \parallel \overline{FG}$. $\angle EKF \cong \angle GKH$ (vertical angles). $\triangle EFK \cong \triangle GHK$ according to SAS. $\angle FEK \cong \angle HGK$ according to the CPCTC. $\angle FEK$ and $\angle HGK$ are alternate interior angles, showing that $\overline{EF} \parallel \overline{GH}$. Since $\overline{EH} \parallel \overline{FG}$ and $\overline{EF} \parallel \overline{GH}$, this shows that EFGH is a parallelogram.

14. Main ideas: $\triangle LMO \cong \triangle MNO$ according to SSS ($\overline{LM} \cong \overline{NO}$ and $\overline{LO} \cong \overline{MN}$ are given and $\overline{MO}$ is shared). $\angle LMO \cong \angle NOM$ according to the CPCTC. $\angle LMO$ and $\angle NOM$ are alternate interior angles, showing that $\overline{LM} \parallel \overline{NO}$. $\angle LOM \cong \angle NMO$ according to the CPCTC. $\angle LOM$ and $\angle NMO$ are alternate interior angles, showing that $\overline{LO} \parallel \overline{MN}$. Since $\overline{LM} \parallel \overline{NO}$ and $\overline{LO} \parallel \overline{MN}$, this shows that LMNO is a parallelogram.

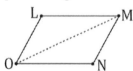

15. Main ideas: $\overline{PQ} \parallel \overline{RS}$ and $\overline{PS} \parallel \overline{QR}$ (definition of parallelogram). According to the parallel postulate (Chapter 1), $\angle QPS + \angle RSP = 180°$ and $\angle PQR + \angle SRQ = 180°$ since $\overline{PQ} \parallel \overline{RS}$, and $\angle SPQ + \angle RQP = 180°$ and $\angle PSR + \angle QRS = 180°$ since $\overline{PS} \parallel \overline{QR}$.

16. Main ideas: $\angle LMN \cong \angle LON$ and $\angle MLO \cong \angle MNO$ (given). The interior angles of a quadrilateral add up to $360°$: $\angle LMN + \angle MNO + \angle LON + \angle MLO = 360°$. Since $\angle LMN$

$\cong \angle$LON and $\angle$MLO $\cong \angle$MNO, this sum simplifies to $2\angle$LMN + $2\angle$MLO = 360°. Divide by 2 on both sides: $\angle$LMN + $\angle$MLO = 180°. $\overline{LO} \parallel \overline{MN}$ according to the parallel postulate (Chapter 1). Plug $\angle$MLO $\cong \angle$MNO into the previous equation: $\angle$LMN + $\angle$MNO = 180°. $\overline{LM} \parallel \overline{NO}$ according to the parallel postulate. Since $\overline{LM} \parallel \overline{NO}$ and $\overline{LO} \parallel \overline{MN}$, this shows that LMNO is a parallelogram.

17. Main ideas: AE = DE = AD/2 and BE = CE = BC/2 because the diagonals of a parallelogram bisect each other's lengths. $\overline{AD} \cong \overline{BC}$ (given). $\overline{AE} \cong \overline{DE} \cong \overline{BE} \cong \overline{CE}$ (combine the first two steps). $\overline{AB} \parallel \overline{CD}$ (definition of parallelogram). $\angle$BAE $\cong \angle$EDC (alternate interior angles). $\angle$EDC $\cong \angle$ECD (since $\triangle$CDE is isosceles, $\overline{CE} \cong \overline{DE}$). $\angle$BAE $\cong \angle$ECD (the transitive property). $\angle$EAC $\cong \angle$ECA (since $\triangle$ACE is isosceles, $\overline{AE} \cong \overline{CE}$). $\angle$BAC + $\angle$ACD = 180° (adjacent angles of a parallelogram are supplementary). $\angle$BAE + $\angle$EAC + $\angle$ECA +$\angle$ECD = 180° (since $\angle$BAC = $\angle$BAE + $\angle$EAC and $\angle$ACD = $\angle$ECA + $\angle$ECD). $2\angle$BAE + $2\angle$EAC = 180° (since $\angle$BAE $\cong \angle$ECD and $\angle$EAC $\cong \angle$ECA). Divide by 2 on both sides: $\angle$BAE + $\angle$EAC = $\angle$BAC = 90°. Similarly, $\angle$ACD = 90°, $\angle$BDC = 90°, and $\angle$ABD = 90°.

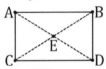

18. Main ideas: Draw altitudes $\overline{RT}$ and $\overline{QU}$ (which are congruent) as shown below to make right triangles. Apply the Pythagorean theorem to $\triangle$PRT: $PR^2 = PT^2 + RT^2$. Apply the Pythagorean theorem to $\triangle$QSU: $QS^2 = QU^2 + SU^2$. Since $\overline{RT} \cong \overline{QU}$, this equation becomes $QS^2 = RT^2 + SU^2$. Add these equations together.

$$PR^2 + QS^2 = PT^2 + 2RT^2 + SU^2$$

Note that PT = PQ − QT = RS − RU (since $\overline{RS} \cong \overline{PQ}$ and $\overline{QT} \cong \overline{RU}$) and SU = RS + RU.

$$PR^2 + QS^2 = (RS - RU)^2 + 2RT^2 + (RS + RU)^2$$

Recall from algebra that $(x - y)^2 = x^2 - 2xy + y^2$ and $(x + y)^2 = x^2 + 2xy + y^2$.

$$PR^2 + QS^2 = RS^2 - 2(RS)(RU) + RU^2 + 2RT^2 + RS^2 + 2(RS)(RU) + RU^2$$
$$PR^2 + QS^2 = 2RS^2 + 2RU^2 + 2RT^2$$

Apply the Pythagorean theorem to $\triangle$QRT: $QT^2 + RT^2 = RU^2 + RT^2 = QR^2$.

$$PR^2 + QS^2 = 2RS^2 + 2QR^2$$

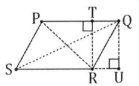

19. Main ideas: $\overline{AB} \cong \overline{BD} \cong \overline{CD} \cong \overline{AC}$ (definition of rhombus). AE = DE = AD/2 and BE = CE = BC/2 because the diagonals of any parallelogram (including a rhombus) bisect each other's lengths. $\triangle ABE \cong \triangle BDE$ according to SSS ($\overline{AB} \cong \overline{BD}$, $\overline{AE} \cong \overline{DE}$, and $\overline{BE}$ is shared). ∠ABE ≅ ∠DBE according to the CPCTC. It can similarly be shown that ∠BAE ≅ ∠CAE, ∠ACE ≅ ∠DCE, and ∠CDE ≅ ∠BDE.

20. Main ideas: $\overline{TU} \cong \overline{TV}$ and $\overline{UW} \cong \overline{VW}$ (definition of kite). $\triangle TUW \cong \triangle TVW$ according to SSS ($\overline{TU} \cong \overline{TV}$, $\overline{UW} \cong \overline{VW}$, and $\overline{TW}$ is shared). ∠UTW ≅ ∠VTW and ∠UWT ≅ ∠VWT according to the CPCTC. Note: $\overline{TW}$ is an angle bisector, but $\overline{UV}$ is NOT.

21. Main ideas: $\overline{TU} \cong \overline{TV}$ and $\overline{UW} \cong \overline{VW}$ (definition of kite). ∠UTW ≅ ∠VTW and ∠UWT ≅ ∠VWT (see the solution to Problem 21). $\triangle UTX \cong \triangle VTX$ according to SAS ($\overline{TU} \cong \overline{TV}$, ∠UTW ≅ ∠VTW, and $\overline{TX}$ is shared). ∠TXV + ∠TXU = 180° (supplementary angles). ∠TXV ≅ ∠TXU according to the CPCTC, such that 2∠TXU = 180°. Divide by 2 on both sides: ∠TXU = 90°.

22. Main ideas: Draw diagonal $\overline{BD}$, as shown below. The area of $\triangle BCD$ equals $\frac{1}{2}(CD)h$ and the area of $\triangle ABD$ equals $\frac{1}{2}(AB)h$. Opposite sides of a parallelogram are congruent: $\overline{AB} \cong \overline{CD}$. The area of parallelogram ABCD is $\frac{1}{2}(CD)h + \frac{1}{2}(AB)h$. Since $\overline{AB} \cong \overline{CD}$, this becomes $\frac{1}{2}(CD)h + \frac{1}{2}(CD)h = (CD)h$, which equals the base times the height.

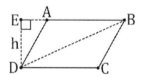

23. Main ideas: $\overline{AD} \perp \overline{BC}$ (the diagonals of a rhombus are perpendicular, as shown in Example 5). Therefore, $\triangle ABC$ has base $\overline{BC}$, altitude $\overline{AE}$, and area $\frac{1}{2}(BC)AE$. Similarly, $\triangle BCD$ has base $\overline{BC}$, altitude $\overline{DE}$, and area $\frac{1}{2}(BC)DE$. Add these areas to find the area of rhombus ABCD: $\frac{1}{2}(BC)AE + \frac{1}{2}(BC)DE$. Since a rhombus is a parallelogram, the diagonals bisect each other's lengths: $\overline{AE} \cong \overline{DE}$. The area of the rhombus is $\frac{1}{2}(BC)AE + \frac{1}{2}(BC)AE$ $= \frac{1}{2}(BC)(2AE)$. Since AD = 2AE, this becomes $\frac{1}{2}(BC)(AD)$, which is half the product of the diagonals. Note: An important point in this proof is that the diagonals of a rhombus are perpendicular. If a general parallelogram is split into triangles by a diagonal, half the other diagonal will NOT be the height of a triangle unless it is a rhombus.

24. Main ideas: Draw altitudes $\overline{AE}$ and $\overline{BF}$ as shown below to divide the trapezoid into two right triangles and a rectangle. Add the areas of the triangles and rectangle to get the area of the trapezoid.

$$A = \frac{1}{2}(DE)(AE) + (AB)(AE) + \frac{1}{2}(CF)(BF)$$

Note that $\overline{BF} \cong \overline{AE}$ and $\overline{AB} \cong \overline{EF}$.

$$A = \frac{1}{2}(DE)(AE) + (AB)(AE) + \frac{1}{2}(CF)(AE)$$

Apply the distributive property to factor AE out.

$$A = \left(\frac{DE}{2} + AB + \frac{CF}{2}\right)(AE) = \left(\frac{DE}{2} + \frac{2AB}{2} + \frac{CF}{2}\right)(AE)$$

$$A = \frac{1}{2}(DE + 2AB + CF)(AE) = \frac{1}{2}(DE + AB + AB + CF)(AE)$$

Recall that $\overline{AB} \cong \overline{EF}$ and note that $DE + EF + CF = CD$.

$$A = \frac{1}{2}(DE + AB + EF + CF)(AE) = \frac{1}{2}(CD + AB)(AE) = \frac{a+b}{2}h$$

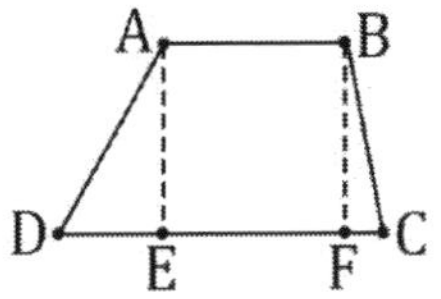

25. Main ideas: $\overline{TW} \perp \overline{UV}$ (the diagonals of a kite are perpendicular; see Problem 21). Therefore, ΔTUV has base $\overline{UV}$, altitude $\overline{TX}$, and area $\frac{1}{2}(UV)TX$. Similarly, ΔUVW has base $\overline{UV}$, altitude $\overline{WX}$, and area $\frac{1}{2}(UV)WX$. The area of kite TUWV is $\frac{1}{2}(UV)TX + \frac{1}{2}(UV)WX = \frac{1}{2}(UV)(TX + WX)$. Since $TX + WX = TW$, the area of the kite is $\frac{1}{2}(UV)(TW)$, which is half the product of the diagonals.

26. (A) Yes. (B) "If a shape is a quadrilateral, it is a parallelogram." The converse is false because there are other kinds of quadrilaterals besides parallelograms, such as trapezoids and kites.

(C) "If a shape is not a parallelogram, it is not a quadrilateral." The inverse is false for the same reason that the converse is false.

(D) "If a shape is not a quadrilateral, it is not a parallelogram." The contrapositive is true.

Chapter 10 Answers

1. (A) 720° (or 4π rad) and 360° (or 2π rad). (B) 900° (or 5π rad) and 360° (or 2π rad). (C) 1620° (or 9π rad) and 360° (or 2π rad). Notes: The sum of the interior angles is 180°(N − 2) in degrees or π(N − 2) in radians. The sum of the exterior angles is 360° in degrees or 2π in radians. For a hexagon, N = 6. For a heptagon, N = 7. For a hendecagon, N = 11. To convert from degrees to radians, multiply by $\frac{\pi}{180°}$.

2. (A) 108° (or $\frac{3\pi}{5}$ rad) and 72° (or $\frac{2\pi}{5}$ rad). (B) 135° (or $\frac{3\pi}{4}$ rad) and 45° (or $\frac{\pi}{4}$ rad). (C) 150° (or $\frac{5\pi}{6}$ rad) and 30° (or $\frac{\pi}{6}$ rad). Notes: Each interior angle equals $180° - \frac{360°}{N}$ in degrees or $\pi - \frac{2\pi}{N}$ in radians. Each exterior angle equals $\frac{360°}{N}$ in degrees or $\frac{2\pi}{N}$ in radians. For a pentagon, N = 5. For an octagon, N = 8. For a dodecagon, N = 12. To convert from degrees to radians, multiply by $\frac{\pi}{180°}$.

3. Yes. N = 4. 90°.

4. N = 17. Check: 180°(17 − 2) = 180°(15) = 2700°.

5. No. The sum of the exterior angles equals 360°, regardless of the number of sides.

6. N = 15. Check: $180° - \frac{360°}{15} = 180° - 24° = 156°$.

7. N = 20. Check: $\frac{360°}{20} = 18°$.

8. 127°. Check: 125° + 106° + 108° + 127° + 127° + 127° = 720°. Notes: The three unknown interior angles are marked as congruent. N = 6 such that 180°(6 − 2) = 180°(4) = 720°.

9. $\theta = 150°$ and $\varphi = 100°$. Check: 150° + 150° + 150° + 150° + 100° + 100° + 100° = 900°. Notes: N = 7 such that 180°(7 − 2) = 180°(5) = 900° and 150°:100° = 3:2.

10. P = 8 and A = $1 + 2\sqrt{2}$. Notes: Two angles are marked as right angles. The other angles are 135° such that 90°×2 + 135°×4 = 720°. N = 6 such that 180°(6 − 2) = 180°(4) = 720°. Draw two vertical lines to divide the hexagon into a rectangle and two 45°-45°-90° triangles, as shown on the following page. Since the ratio of the sides of a 45°-45°-90° triangle is 1:1:$\sqrt{2}$, the vertical line segments are $\sqrt{2}$. The perimeter is P = 2(2) + 1(4) = 8. The area of each triangle is $\frac{1}{2}(1)(1) = \frac{1}{2}$ since one leg is the base

and the other leg is the altitude of the triangle (since the two legs are perpendicular). The area of the rectangle is $2\sqrt{2}$. The total area is $\frac{1}{2} + \frac{1}{2} + 2\sqrt{2} = 1 + 2\sqrt{2}$.

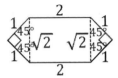

11. $d = 2$, $\alpha = 60°$, $P = 12 + 2\sqrt{3}$, and $A = 7\sqrt{3}$. Notes: $\alpha = 60°$ since the interior angles of a pentagon add up to $180°(5 - 2) = 180°(3) = 540°$. Draw one horizontal line and two vertical lines to divide the pentagon into a rectangle, two 30°-60°-90° triangles, and one 30°-30°-120° triangle, as shown below. The sides of a 30°-30°-120° triangle come in the ratio $1{:}1{:}\sqrt{3}$ (recall Problem 12 from Chapter 5), such that its base has a length of $4\sqrt{3}$. Since the rectangle has a length of $2\sqrt{3}$, each 30°-60°-90° triangle must have a base of $\sqrt{3}$ (so that $\sqrt{3} + 2\sqrt{3} + \sqrt{3} = 4\sqrt{3}$). Since the sides of a 30°-60°-90° triangle come in the ratio $1{:}\sqrt{3}{:}2$, each 30°-60°-90° triangle has a height of 1, a base of $\sqrt{3}$, and a hypotenuse of 2. The perimeter is $P = 4 + 4 + 2 + 2 + 2\sqrt{3} = 12 + 2\sqrt{3}$ $= 2(6 + \sqrt{3})$. The area of the rectangle is $2\sqrt{3}$. The area of each 30°-60°-90° triangle is $\frac{1}{2}\sqrt{3}(1) = \frac{\sqrt{3}}{2}$. The 30°-30°-120° triangle has a height of 2 (which can be seen by splitting it into two 30°-60°-90° triangles with hypotenuses of 4). The area of the 30°-30°-120° triangle is $\frac{1}{2}(2)(4\sqrt{3}) = 4\sqrt{3}$. Add the areas together to find the total area:

$2\sqrt{3} + \frac{\sqrt{3}}{2} + \frac{\sqrt{3}}{2} + 4\sqrt{3} = 2\sqrt{3} + \sqrt{3} + 4\sqrt{3} = 7\sqrt{3}$.

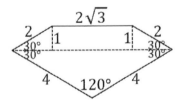

12. $P = 8L$, $A = 2L^2 + 2L^2\sqrt{2}$ (which is equivalent to $2L^2 + \frac{4L^2}{\sqrt{2}}$ since $\frac{4L^2}{\sqrt{2}}\frac{\sqrt{2}}{\sqrt{2}} = \frac{4L^2\sqrt{2}}{2} =$ $2L^2\sqrt{2}$), and the apothem equals $\frac{L}{2} + \frac{L\sqrt{2}}{2}$ (which is equivalent to $\frac{L}{2} + \frac{L}{\sqrt{2}}$ since $\frac{L}{\sqrt{2}}\frac{\sqrt{2}}{\sqrt{2}} = \frac{L\sqrt{2}}{2}$). Notes: One way to find the area is to extend the horizontal and vertical sides to make a large square, as illustrated on the following page. The area of the octagon equals the area of the square minus the area of four 45°-45°-90° triangles. The hypotenuse of

each 45°-45°-90° triangle equals L. Since the ratio of the sides of a 45°-45°-90° triangle is 1:1:$\sqrt{2}$, the length of each leg equals $\frac{L}{\sqrt{2}}$, which is equivalent to $\frac{L}{\sqrt{2}} = \frac{L}{\sqrt{2}} \frac{\sqrt{2}}{\sqrt{2}} = \frac{L\sqrt{2}}{2}$. The ratio $\frac{L}{\sqrt{2}}:\frac{L}{\sqrt{2}}$:L is equivalent to 1:1:$\sqrt{2}$ (to see this, multiply 1:1:$\sqrt{2}$ by L and divide by $\sqrt{2}$).

The large square has edge length equal to $\frac{L}{\sqrt{2}} + L + \frac{L}{\sqrt{2}} = L + \frac{2L}{\sqrt{2}}$, which is equivalent to L + L$\sqrt{2}$ since $\frac{2L}{\sqrt{2}} = \frac{2L}{\sqrt{2}} \frac{\sqrt{2}}{\sqrt{2}} = L\sqrt{2}$. The area of the square equals:

$$\left(L + L\sqrt{2}\right)^2 = L^2 + 2L^2\sqrt{2} + 2L^2 = 3L^2 + 2L^2\sqrt{2}$$

Recall from algebra that $(x + y)^2 = x^2 + 2xy + y^2$ and that $\left(\sqrt{2}\right)^2 = 2$. The area of the square is equivalent to $3L^2 + \frac{4L^2}{\sqrt{2}}$ since $\frac{4L^2}{\sqrt{2}} = \frac{4L^2}{\sqrt{2}} \frac{\sqrt{2}}{\sqrt{2}} = \frac{4L^2\sqrt{2}}{2} = 2L^2\sqrt{2}$. The area of each 45°-45°-90° triangle is $\frac{1}{2}\left(\frac{L}{\sqrt{2}}\right)\left(\frac{L}{\sqrt{2}}\right) = \frac{L^2}{4}$. The area of the octagon is:

$$3L^2 + 2L^2\sqrt{2} - 4\frac{L^2}{4} = 3L^2 + 2L^2\sqrt{2} - L^2 = 2L^2 + 2L^2\sqrt{2}$$

The area may be factored as $2L^2\left(1 + \sqrt{2}\right)$. The area is equivalent to $2L^2 + \frac{4L^2}{\sqrt{2}}$, but the preferred form is $2L^2 + 2L^2\sqrt{2}$ or $2L^2\left(1 + \sqrt{2}\right)$ because the denominator is rational in these forms. The apothem equals one-half the edge length of the square: $\frac{1}{2}\left(L + L\sqrt{2}\right)$ $= \frac{L}{2} + \frac{L\sqrt{2}}{2} = \frac{L}{2}\left(1 + \sqrt{2}\right)$, which is equivalent to $\frac{L}{2} + \frac{L}{\sqrt{2}}$ since $\frac{L}{\sqrt{2}} = \frac{L}{\sqrt{2}} \frac{\sqrt{2}}{\sqrt{2}} = \frac{L\sqrt{2}}{2}$.

13. $\theta = 36°$ and $\varphi = 252°$. Notes: Each interior angle of the regular pentagon is $\alpha = 180° - \frac{360°}{5} = 180° - 72° = 108°$. In the right diagram on the next page, α and γ form two pairs of vertical angles. Since these two pairs of angles form a full circle, $2\alpha + 2\gamma = 360°$. Plug $\alpha = 108°$ into this equation:

$$2(108°) + 2\gamma = 360°$$
$$216° + 2\gamma = 360°$$
$$2\gamma = 360° - 216°$$
$$2\gamma = 144°$$
$$\gamma = 72°$$

Each triangle with θ is an isosceles triangle. The interior angles of a triangle add up to: $\theta + \gamma + \gamma = 180°$. Plug $\gamma = 72°$ into this equation:

$$\theta + 72° + 72° = 180°$$
$$\theta = 180° - 144° = 36°$$

Compare the two diagrams below to see that $\varphi = \alpha + \gamma + \gamma$. Plug $\alpha = 108°$ and $\gamma = 72°$ into this equation: $\varphi = 108° + 72° + 72° = 252°$. Check the answers: The interior angles of a decagon add up to $180°(10-2) = 180°(8) = 1440°$, which agrees with $5\theta + 5\varphi = 5(36°) + 5(252°) = 180° + 1260° = 1440°$.

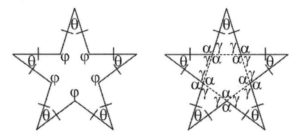

14. $d = \frac{\sqrt{3}}{3}$ (which is equivalent to $\frac{1}{\sqrt{3}}$), $\angle 1 = 240°$, $P = 2 + \frac{2\sqrt{3}}{3}$ (which is equivalent to $2 + \frac{2}{\sqrt{3}}$), and $A = \frac{\sqrt{3}}{6}$. Notes: Draw a horizontal line at the top of the concave kite to make an equilateral triangle. See the next page. The horizontal line segment is 1 unit long. $\angle 1 = 240°$ so that the four interior angles of the kite add up to $30° + 30° + 60° + 240° = 360°$. The small triangle at the top is a 30°-30°-120° triangle. The sides of a 30°-30°-120° triangle come in the ratio $1:1:\sqrt{3}$ (recall Problem 12 from Chapter 5), such that each of the top sides of the concave kite has a length of $\frac{1}{\sqrt{3}}$, which is equivalent to $\frac{\sqrt{3}}{3}$ since $\frac{1}{\sqrt{3}}\frac{\sqrt{3}}{\sqrt{3}} = \frac{\sqrt{3}}{3}$. The ratio $\frac{1}{\sqrt{3}}:\frac{1}{\sqrt{3}}:1$ is equivalent to $1:1:\sqrt{3}$ (to see this, multiply $\frac{1}{\sqrt{3}}:\frac{1}{\sqrt{3}}:1$ by $\sqrt{3}$). The height of the 30°-30°-120° triangle is $\frac{1}{2\sqrt{3}}$, which is equivalent to $\frac{\sqrt{3}}{6}$ since $\frac{1}{2\sqrt{3}} = \frac{1}{2\sqrt{3}}\frac{\sqrt{3}}{\sqrt{3}} = \frac{\sqrt{3}}{2(3)} = \frac{\sqrt{3}}{6}$. (To see this, divide the 30°-30°-120° triangle into two 30°-60°-90° triangles.) The area of the 30°-30°-120° triangle is $\frac{1}{2}(1)\left(\frac{\sqrt{3}}{6}\right) = \frac{\sqrt{3}}{12}$. The area of the equilateral triangle is $\frac{\sqrt{3}}{4}$ (as discussed in Chapter 6). Subtract the area of the 30°-30°-120° triangle from the area of the equilateral triangle in order to find the area of the concave kite.

$$\frac{\sqrt{3}}{4} - \frac{\sqrt{3}}{12} = \frac{3\sqrt{3}}{12} - \frac{\sqrt{3}}{12} = \frac{2\sqrt{3}}{12} = \frac{\sqrt{3}}{6}$$

Note that $\frac{\sqrt{3}}{6}$ is equivalent to $\frac{1}{2\sqrt{3}}$ since $\frac{1}{2\sqrt{3}}\frac{\sqrt{3}}{\sqrt{3}} = \frac{\sqrt{3}}{2(3)} = \frac{\sqrt{3}}{6}$. Since each of the top sides of the concave kite has a length of $\frac{\sqrt{3}}{3}$ (equivalent to $\frac{1}{\sqrt{3}}$), the perimeter of the concave kite is $P = 2 + \frac{2\sqrt{3}}{3} = \frac{6+2\sqrt{3}}{3} = 2\left(1 + \frac{\sqrt{3}}{3}\right) = 2\left(\frac{3+\sqrt{3}}{3}\right)$, which is equivalent to $2 + \frac{2}{\sqrt{3}}$.

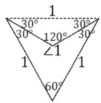

15. Main ideas: Angles α, β, γ, and so on, ending with η, form a full circle. (As explained in the problem, there are additional dashed lines extending from point P to vertices that have not been drawn. An unknow number of angles between ε and η are not shown.) Since these angles form a full circle, they add up to $360°$. The three dots (…) below mean "and so on."

$$\alpha + \beta + \gamma + \delta + \varepsilon + \cdots + \eta = 360°$$

The interior angles of each triangle add up to $180°$, such that $\alpha + \angle 1a + \angle 1b = 180°$, $\beta + \angle 2a + \angle 2b = 180°$, $\gamma + \angle 3a + \angle 3b = 180°$, etc., up to $\eta + \angle Na + \angle Nb = 180°$. In the equation above, $\alpha = 180° - \angle 1a - \angle 1b$, $\beta = 180° - \angle 2a - \angle 2b$, and so on.

$$180° - \angle 1a - \angle 1b + 180° - \angle 2a - \angle 2b + 180° - \angle 3a - \angle 3b + 180° - \angle 4a - \angle 4b$$
$$+ 180° - \angle 5a - \angle 5b + \cdots + 180° - \angle Na - \angle Nb = 360°$$

Observe that $\angle 1b + \angle 2a = \angle 1$, $\angle 2b + \angle 3a = \angle 2$, $\angle 3b + \angle 4a = \angle 3$, $\angle 4b + \angle 5a = \angle 4$, up to $\angle Nb + \angle 1a = \angle N$. The $180°$ appears N times on the left-hand side.

$$180°N - \angle 1 - \angle 2 - \angle 3 - \angle 4 - \angle 5 - \cdots - \angle N = 360°$$

Add the interior angles to both sides and subtract $360°$ from both sides.

$$180°N - 360° = \angle 1 + \angle 2 + \angle 3 + \angle 4 + \angle 5 + \cdots + \angle N$$
$$180°(N - 2) = \angle 1 + \angle 2 + \angle 3 + \angle 4 + \angle 5 + \cdots + \angle N$$

16. Main ideas: As shown in the solution to Problem 15, the sum of the interior angles for any polygon equals $180°N - 360°$.

$$\angle 1 + \angle 2 + \angle 3 + \cdots + \angle N = 180°N - 360°$$

For an equiangular polygon, the interior angles are congruent. Set the N angles on the left-hand side equal. There are N interior angles added together.

$$N\alpha = 180°N - 360°$$

Divide by N on each side of the equation.

$$\alpha = \frac{180°N - 360°}{N} = \frac{180°N}{N} - \frac{360°}{N} = 180° - \frac{360°}{N}$$

17. (A) Yes. The statement is true because a polygon must be both equiangular and equilateral in order to be regular.

(B) "If a polygon is not regular, it is not equiangular." The converse is false because it is possible for a polygon to be equiangular, but not equilateral. A polygon that is only equiangular (and not equilateral) is not regular.

(C) "If a polygon is equiangular, it is regular." The inverse is false for the same reason that the converse is false. A polygon that is equiangular is not necessarily equilateral, but must be both equiangular and equilateral in order to be regular.

(D) "If a polygon is regular, it is equiangular." The contrapositive is true. A polygon that is regular is both equiangular and equilateral.

Glossary

AA: angle-angle. If two interior angles are congruent for two triangles, the triangles are similar.

AAS: angle-angle-side. If two interior angles and a side that is not between those angles are congruent for two triangles, the triangles are congruent.

acute angle: an angle that is less than 90°.

acute triangle: a triangle with three acute interior angles.

alternate exterior angles: ∠1 and ∠8 below, and ∠2 and ∠7.

alternate interior angles: ∠3 and ∠6 below, and ∠4 and ∠5.

$$\angle1/\angle2$$
$$A \quad \angle3/\angle4 \quad B$$
$$C \angle5/\angle6 \quad D$$
$$\angle7/\angle8$$

altitude: a line segment that is perpendicular to one side of a triangle and connects to the vertex opposite to that side. When the side it connects to is the base of the triangle, the altitude is the height.

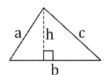

angle: the shape made when two line segments are joined at a common endpoint.

angle bisector: a line segment that cuts an angle into two smaller congruent angles.

apothem: the distance from the center of a regular polygon to the midpoint of one of its sides.

ASA: angle-side-angle. If two interior angles and a side that is between those angles are congruent for two triangles, the triangles are congruent.

askew: two lines (or line segments) that are not parallel, but which do not intersect. If two (infinite) lines are askew, they do not lie in the same plane.

axiom: a basic principle that is considered to be self-evident.

base: the bottom side of a triangle, or either of the parallel sides of a trapezoid.

bisect: cut a shape into two congruent parts.

bisector: a line that bisects a shape. See angle bisector and perpendicular bisector.

centroid: the point where the three medians of a triangle intersect.

circle: a curve for which every point on the curve is equidistant from its center.

circumcenter: the point where the three perpendicular bisectors of a triangle intersect.

circumradius: the distance from the circumcenter to any vertex of a triangle.

coincident: occupying the same position in space.

collinear: lying on the same line.

complementary angles: two angles that together form a 90° angle.

concave: a polygon with at least one interior angle that is greater than 180°.

concurrent: three or more lines that intersect at a single point.

congruent: having the same shape and size (even if one shape is rotated relative to the other, and even if one shape is the mirror image of the other).

contrapositive: when a statement of the form, "If A, then B," is put in the form, "If not B, then not A."

converse: when a statement of the form, "If A, then B," is put in the form, "If B, then A."

convex: a polygon where every interior angle is less than 180°.

coplanar: lying within the same plane.

corollary: a mathematical statement that can be obtained from a theorem with very little effort.

corresponding angles: ∠1 and ∠5 below, ∠2 and ∠6, ∠3 and ∠7, and ∠4 and ∠8.

CPCTC: corresponding parts of congruent triangles are congruent.

decagon: a polygon with ten sides.

degree: 1/360th of a circle, as measured from the center.

diagonal: a line segment that joins two non-adjacent vertices in a polygon.

dimension: a measure of extent.

dodecagon: a polygon with twelve sides.

edge: a side of a polygon.

endpoints: the points at the end of a line segment.

equiangular: a polygon where all of the interior angles are congruent.

equidistant: equally distant.

equilateral: a polygon where all of the sides are congruent.

exterior angle: an angle that is formed by one side of a polygon and a line that extends from an adjacent side, such as φ in the diagram below.

height: a line segment that is perpendicular to the base of a triangle and connects to the vertex opposite to the base.

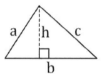

hendecagon: a polygon with eleven sides.

heptagon: a polygon with seven sides.

Heron's formula: an equation to find the area of a triangle in terms of the lengths of the three sides, where $P = a + b + c$ is the perimeter.

$$A = \sqrt{\frac{P}{2}\left(\frac{P}{2} - a\right)\left(\frac{P}{2} - b\right)\left(\frac{P}{2} - c\right)}$$

hexagon: a polygon with six sides.

horizontal line: a line that runs across to the left and right.

hypotenuse: the longest side of a right triangle, which is opposite to the 90° angle.

incenter: the point where the three angle bisectors of a triangle intersect.

inradius: the shortest distance from the incenter to any side of a triangle.

interior angle: an angle that is formed by two sides that meet at a vertex of a polygon, and which lies inside of the polygon.

intersect: when two lines (or line segments or rays) cross paths.

inverse: when a statement of the form, "If A, then B," is put in the form, "If not A, then not B."

isosceles: having two congruent sides.

kite: a quadrilateral with no parallel sides that has two pairs of congruent sides.

leg: either of the shorter sides of a right triangle, which touch the 90° angle, or the non-parallel sides of a trapezoid.

lemma: a theorem which, once it is proven, helps to prove a more significant theorem.

LH: leg-hypotenuse. If the hypotenuse and one leg of two right triangles are congruent, the triangles are congruent.

line: a straight path that is infinite in each direction.

line segment: a straight, finite path that connects two endpoints.

median: a line segment that joins one vertex to the midpoint of the opposite side of a triangle, or a line segment that joins the midpoints of the legs of a trapezoid.

midline: a midsegment.

midpoint: a point that bisects a line segment.

midsegment: a line segment that connects the midpoints of two sides of a triangle.

nonagon: a polygon with nine sides.

obtuse angle: an angle that is greater than 90°.

obtuse triangle: a triangle with one obtuse angle and two acute angles.

octagon: a polygon with eight sides.

orthocenter: the point where the three altitudes of a triangle intersect.

parallel: lines that extend in the same direction and are the same distance apart at any position such that they will never intersect.

parallel postulate: if the same-side interior angles do not add up to exactly 180°, then lines $\overleftrightarrow{AB}$ and $\overleftrightarrow{CD}$ intersect (and therefore are not parallel), but if the same-side interior angles (either $\angle 3 + \angle 5$ or $\angle 4 + \angle 6$) do add up to exactly 180°, then lines $\overleftrightarrow{AB}$ and $\overleftrightarrow{CD}$ must be parallel (in this case, $\overleftrightarrow{AB}$ and $\overleftrightarrow{CD}$ do not intersect).

$$\begin{array}{ccc} & \angle 1 / \angle 2 & \\ A & \angle 3 / \angle 4 & B \\ C & \angle 5 / \angle 6 & D \\ & \angle 7 / \angle 8 & \end{array}$$

parallelogram: a quadrilateral with two pairs of opposite sides that are parallel and congruent.

pentagon: a polygon with five sides.

perimeter: the total distance around the edges of a polygon. The sum of the side lengths.

perpendicular: lines (or line segments or rays) that meet at a 90° angle.

perpendicular bisector: a line that passes perpendicularly through a line segment and cuts the line segment in half.

pi: the ratio of the circumference of any circle to its diameter, which approximately equals 3.14159 (with the digits continuing forever without repeating).

plane: a flat surface that extends infinitely in two independent directions.

polygon: a closed plane figure that is bounded by straight sides.

postulate: a basic principle that is considered to be self-evident.

Pythagorean theorem: the sum of the squares of the lengths of the two legs of a right triangle equals the square of the hypotenuse. For the right triangle below, $a^2 + b^2 = c^2$.

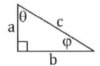

quadrilateral: a polygon with four sides.

radian: a unit of angular measure equivalent to $\frac{180°}{\pi}$, such that the angular measure of one full circle equals 2π radians (equivalent to 360°).

ray: a semi-infinite straight path, extending infinitely in one direction.

rectangle: an equiangular parallelogram. Every interior angle is 90°.

reflex angle: an angle greater than 180°.

reflexive property: two figures that are completely coincident are congruent.

regular: both equilateral and equiangular.

rhombus: an equilateral parallelogram.

right angle: an angle measuring exactly 90°.

right triangle: a triangle with one right angle and two acute angles.

same-side exterior angles: ∠1 and ∠7 below, and ∠2 and ∠8.

same-side interior angles: ∠3 and ∠5 below, and ∠4 and ∠6.

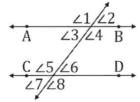

SAS: side-angle-side. If two sides and the interior angle that is formed by those sides are congruent for two triangles, the triangles are congruent.

scalene: a triangle that does not have any two sides with the same length.

side: a line segment that joins two adjacent vertices of a polygon. The sides of a polygon form its boundary.

similar: having the same shape, but not the same size.

square: a regular quadrilateral. It is both equilateral and equiangular. Like a rectangle, every interior angle is 90°.

SSS: side-side-side. If every side is congruent for two triangles, the triangles are congruent.

straight angle: an angle measuring exactly 180°, such as ∠ABC below.

$$A\bullet\!\!\!\rule{5em}{0.4pt}\!\!\!\overset{\textstyle B}{\bullet}\!\!\!\rule{5em}{0.4pt}\!\!\!\bullet C$$

supplementary angles: two angles that together form a 180° angle.

theorem: a mathematical statement that can be proven using postulates (or other theorems that are already known to be true) by applying logical reasoning.

tick mark: a short line drawn through two or more line segments to indicate that the line segments are congruent.

transitive rule: if a equals b and b equals c, it follows that a equals c. The transitive rule may also be applied to demonstrate the congruence of geometric figures.

transversal: a line that intersects at least two other lines.

trapezoid: a quadrilateral with one pair of parallel sides.

triangle: a polygon with three sides.

triangle bisector theorem: an angle bisector divides the opposite side of the triangle into segments in proportion to the lengths of the other two sides. For example, in the diagram below, $\frac{BD}{AB} = \frac{CD}{AC}$ (which is equivalent to $\frac{BD}{CD} = \frac{AB}{AC}$).

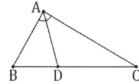

triangle inequality: the sum of the lengths of any two sides of a triangle is greater than the length of the remaining side.

undecagon: a hendecagon.

vertex: the point where the sides of an angle intersect. The plural form is vertices.

vertical angles: angles that appear on opposite sides of a vertex where two lines intersect.

vertical line: a line that runs up and down without any slant.

Index

Notation and Symbols

$\overset{A}{\cdot}$	a single point labeled A
AB	the straight-line distance between points A and B
$\overline{AB}$	a finite line segment connecting point A to point B
$\overleftrightarrow{AB}$	an infinite line passing through points A and B
$\overrightarrow{AB}$	a semi-infinite ray starting at A and passing through B and beyond
$\overleftarrow{AB}$	a semi-infinite ray starting at B and passing through A and beyond
ABC	an infinite plane containing points A, B, and C
$\triangle ABC$	a (finite) triangle with vertices at points A, B, and C
$\angle ABC$	an angle with point B at the vertex, with sides $\overline{AB}$ and $\overline{BC}$
$\angle 1$	a numbered angle
α	an angle indicated by using a lowercase Greek letter
ABCD	a quadrilateral with vertices at points A, B, C, and D
$^\circ$	degrees
rad	radians
π	the constant pi, which is approximately equal to 3.14159
$\perp$	perpendicular to
$\parallel$	parallel to
$\nparallel$	not parallel to
$=$	is equal to
$\cong$	is congruent with
$\sim$	is similar to
a:b	the ratio of a to b

A	area (or a point labeled A)
P	perimeter (or a point labeled P)
b	base
b_1, b_2	the two bases of a trapezoid (these are the two parallel sides)
h	height
d_1, d_2	two diagonals of a polygon (or two distances)
m	the median of a trapezoid (or the slope of a line)
C	the centroid of a triangle (or a point labeled C)
I	the incenter of a triangle (or a point labeled I)
O	the circumcenter of a triangle (or a point labeled O)
H	the orthocenter of a triangle (or a point labeled H)
r	inradius (or the radius of a circle)

Greek Alphabet

α	lowercase alpha	A	uppercase alpha
β	lowercase beta	B	uppercase beta
γ	lowercase gamma	Γ	uppercase gamma
δ	lowercase delta	Δ	uppercase delta
ε	lowercase epsilon	E	uppercase epsilon
ϵ	a common variation of lowercase epsilon		
ζ	lowercase zeta	Z	uppercase zeta
η	lowercase eta	H	uppercase eta
θ	lowercase theta	Θ	uppercase theta
ι	lowercase iota	I	uppercase iota
κ	lowercase kappa	K	uppercase kappa
λ	lowercase lambda	Λ	uppercase lambda
μ	lowercase mu	M	uppercase mu
ν	lowercase nu	N	uppercase nu
ξ	lowercase xi	Ξ	uppercase xi
ο	lowercase omicron	O	uppercase omicron
π	lowercase pi	Π	uppercase pi
ρ	lowercase rho	P	uppercase rho
σ	lowercase sigma	Σ	uppercase sigma
τ	lowercase tau	T	uppercase tau
υ	lowercase upsilon	Υ	uppercase upsilon
φ	lowercase phi	Φ	uppercase phi
ϕ	a common variation of lowercase phi		
χ	lowercase chi	X	uppercase chi
ψ	lowercase psi	Ψ	uppercase psi
ω	lowercase omega	Ω	uppercase omega

Notes:

- ε and ϵ are two common variations of lowercase epsilon.
- φ and ϕ are two common variations of lowercase phi.
- xi (ξ and Ξ) is pronounced zi or si (with a long i sound, like "lie") in English. In Greek, it is pronounced ksee (with a long e sound, like "bee"). However, there are math and science teachers who speak English who pronounce xi as ksee.
- Ξ is uppercase xi, whereas ≡ means "is defined as." The middle line is shorter in Ξ, whereas all three bars have equal length in ≡.
- phi (φ and Φ) and chi (χ and X) may be pronounced in English with a long e (like "bee") or a long i (like "lie"). Although the long i may make more sense from the perspective that many ancient Greek and Latin words in the English language ending with "i" tend to end with a long i (like "lie") sound, such as "cacti" (the plural form of cactus), "octopi," and "alibi," the long e (like "bee") sound preserves the original pronunciation. Both pronunciations are in common use today. In contrast, pi (π and Π) is only pronounced with a long i (like "lie").
- psi (ψ and Ψ) is usually pronounced like the word "sigh" in English. This letter is common in quantum mechanics.
- π (lowercase pi) is usually reserved for the constant that approximately equals 3.14159 (but which continues forever without repeating). It equals the ratio of the circumference of any circle to its diameter.
- θ (lowercase theta) and φ (lowercase phi) are commonly used for angles in many applications of geometry (such as physics and engineering).
- δ (lowercase delta) and ε (lowercase epsilon) are commonly used in calculus.
- ϵ (a variation of lowercase epsilon) is common in set theory notation.
- Δ (uppercase delta) often means "change in." Example: ΔU = the change in U.
- Σ (uppercase sigma) is commonly used as a summation symbol.
- σ (lowercase sigma) is frequently used in statistics.
- Ω (uppercase omega) is the SI unit of resistance (the "Ohm").
- Some Greek letters (such as A, o, and N) are best avoided since they resemble letters in the English alphabet.

The Learning Continues in Volume 2

Circles, Chords, Secants, and Tangents

Coming in the Spring of 2021

WAS THIS BOOK HELPFUL?

Much effort and thought was put into this book, such as:
- Providing more than just the answers in the answer key.
- Careful selection of examples and problems for their instructional value.
- Numerous detailed illustrations to help visualize the principles.
- Coverage of a variety of essential geometry topics.
- A concise review of relevant concepts at the beginning of each chapter.

If you appreciate the effort that went into making this book possible, there is a simple way that you could show it:

Please take a moment to post an honest review.

For example, you can review this book at Amazon.com or Goodreads.com.

Even a short review can be helpful and will be much appreciated. If you are not sure what to write, following are a few ideas, though it is best to describe what is important to you.
- How much did you learn from reading and using this workbook?
- Was the information in the answer key helpful?
- Were you able to understand the ideas?
- Was it helpful to follow the examples while solving the problems?
- Would you recommend this book to others? If so, why?

Do you believe that you found a mistake? Please email the author, Chris McMullen, at greekphysics@yahoo.com to ask about it. One of two things will happen:
- You might discover that it wasn't a mistake after all and learn why.
- You might be right, in which case the author will be grateful and future readers will benefit from the correction. Everyone is human.

ABOUT THE AUTHOR

Dr. Chris McMullen has over 20 years of experience teaching university physics in California, Oklahoma, Pennsylvania, and Louisiana. Dr. McMullen is also an author of math and science workbooks. Whether in the classroom or as a writer, Dr. McMullen loves sharing knowledge and the art of motivating and engaging students.

The author earned his Ph.D. in phenomenological high-energy physics (particle physics) from Oklahoma State University in 2002. Originally from California, Chris McMullen earned his Master's degree from California State University, Northridge, where his thesis was in the field of electron spin resonance.

As a physics teacher, Dr. McMullen observed that many students lack fluency in fundamental math skills. In an effort to help students of all ages and levels master basic math skills, he published a series of math workbooks on arithmetic, fractions, long division, word problems, algebra, geometry, trigonometry, logarithms, and calculus entitled *Improve Your Math Fluency*. Dr. McMullen has also published a variety of science books, including astronomy, chemistry, and physics workbooks.

Author, Chris McMullen, Ph.D.

MATH

This series of math workbooks is geared toward practicing essential math skills:

- Prealgebra
- Algebra
- Geometry
- Trigonometry
- Logarithms and exponentials
- Calculus
- Fractions, decimals, and percentages
- Long division
- Arithmetic
- Word problems
- Roman numerals
- The four-color theorem and basic graph theory

www.improveyourmathfluency.com

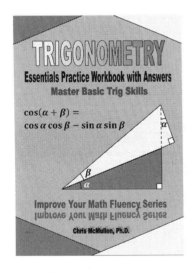

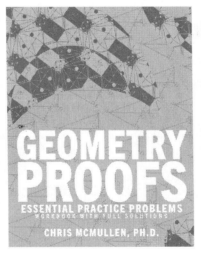

PUZZLES

The author of this book, Chris McMullen, enjoys solving puzzles. His favorite puzzle is Kakuro (kind of like a cross between crossword puzzles and Sudoku). He once taught a three-week summer course on puzzles. If you enjoy mathematical pattern puzzles, you might appreciate:

300+ Mathematical Pattern Puzzles

Number Pattern Recognition & Reasoning

- Pattern recognition
- Visual discrimination
- Analytical skills
- Logic and reasoning
- Analogies
- Mathematics

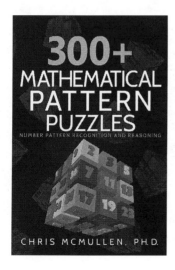

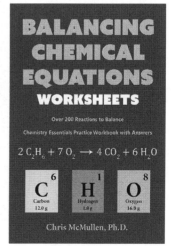

THE FOURTH DIMENSION

Are you curious about a possible fourth dimension of space?

- Explore the world of hypercubes and hyperspheres.
- Imagine living in a two-dimensional world.
- Try to understand the fourth dimension by analogy.
- Several illustrations help to try to visualize a fourth dimension of space.
- Investigate hypercube patterns.
- What would it be like to be a 4D being living in a 4D world?
- Learn about the physics of a possible four-dimensional universe.

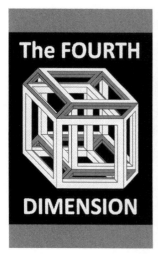

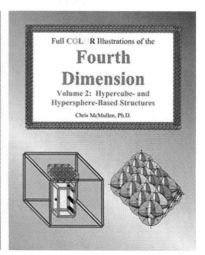

SCIENCE

Dr. McMullen has published a variety of **science** books, including:

- Basic astronomy concepts
- Basic chemistry concepts
- Balancing chemical reactions
- Calculus-based physics textbooks
- Calculus-based physics workbooks
- Calculus-based physics examples
- Trig-based physics workbooks
- Trig-based physics examples
- Creative physics problems
- Modern physics

www.monkeyphysicsblog.wordpress.com

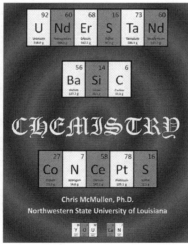
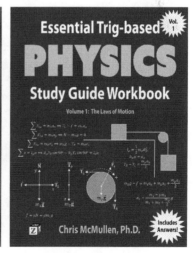

The **FOUR-COLOR THEOREM**
and Basic **GRAPH THEORY**

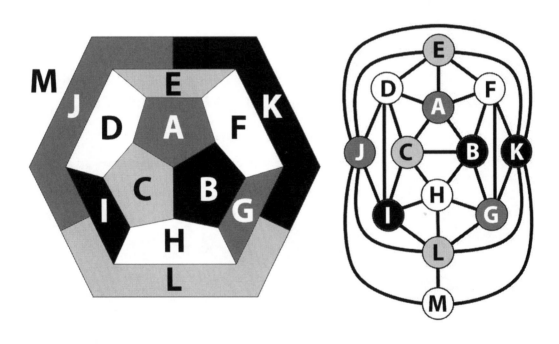

Chris McMullen, Ph.D.

86498869R00116